GLOBAL WARMING AND ECONOMIC DEVELOPMENT

Advances in Computational Economics

VOLUME 2

The titles published in this series are listed at the end of this volume.

Global Warming
and
Economic Development

A Holistic Approach to International Policy Co-operation and Co-ordination

by

Anantha K. Duraiappah
European University Institute,
Florence, Italy

SPRINGER-SCIENCE+BUSINESS MEDIA, B.V.

Library of Congress Cataloging-in-Publication Data

Duraiappah, Anantha K.
Global warming and economic development : a holistic approach to international policy co-operation and co-ordination / by Anantha K. Duraiappah.
p. cm. -- (Advances in computational economics ; v. 2)
Based on the author's thesis (doctoral)--University of Texas, Austin.
Includes bibliographical references and index.
ISBN 978-94-010-4771-5 ISBN 978-94-011-1757-9 (eBook)
DOI 10.1007/978-94-011-1757-9
1. Environmental policy--Economic aspects--Mathematical models.
2. Environmental policy--International cooperation--Mathematical models. 3. Global warming--Economic aspects--Mathematical models.
4. Economic development--Environmental aspects--Mathematical models.
I. Title. II. Series.
HC79.E5D86 1993
338.9--dc20 92-47021

ISBN 978-94-010-4771-5

Printed on acid-free paper

To my grandparents, parents, and wife

Table Of Contents

Chapter Three: The Mathematical Model

List of Tables

Chapter Four

Chapter Five

Chapter Six

Appendix TWO

List of Figures

Preface

The computer revolution both in the hardware as well as in software has made it possible for economists to analyze complex issues which could not be solved in the past by analytical methods. A large library of numerical techniques are now available to economists for solving models ranging from a simple system of linear simultaneous equations to large non-linear dynamic optimization models. We attempt to take advantage of these advancements in computational economics to address the issue of global warming and economic development. The use of computer simulation models has enhanced the understanding of some of the underlying issues in the global warming literature which would have been impossible without these models.

However, to date, the global warming issue has been addressed in a partial equilibrium framework. In other words, the climate scientists tend to specify economic variables as exogenous variables in their global warming models while the economists do the same by specifying the climate variables as exogenous variables in their global warming models. Both approaches ignore important feedback relationships which will be triggered when either economic or climate variables are perturbed. The ideal model structure would be one in which both systems are incorporated within one framework with emphasis on the long run effects of greenhouse gas curbing policies and the corresponding effect on the economic growth potential of the economies.

The task is not easy as the global warming problem is a complex and difficult problem. Let us demonstrate the complex nature by giving a brief overview of the necessary components required for an ideal model structure. First, a global economic system has to be specified. Next, a climate system has to be specified. Third, links between the two systems have to be formulated. The link between the economic and climate system via emission equations can be relatively easily done. However, linking the climate system back to the economic system through damage equations is far more difficult. These equations tell us the impact climate change will have on the economic system. All the equations above will need to be specified as differential or difference equations as both systems are time dependent and this becomes especially crucial in the global warming problem because of the long time lags between emissions and impacts. To add to this complexity, we should then take into account of the large degree of uncertainty inherent in both systems as well as in the linkage equations. And finally, the political issues of the global warming issue have to be explicitly captured as different regions will have different attitudes towards the problem.

By formulating a model in the structure specified above, the following three fundamental questions in the global warming problem can be answered:

- is a reduction in greenhouse gas emissions necessary?
- if the answer is yes to question one, then what is the optimal or sustainable level of reduction?
- Which country or group of countries should be held responsible for achieving the optimal or sustainable level of emissions?

We believe that this book's main contribution to the global warming literature is the description of the methodology used in the formulation of such an ideal model with emphasis on the damage equations. We call the model a holistic optimal growth model for reasons which will become obvious as the reader proceeds through the book.

The book is primarily intended for economists working in the environmental field as the modeling methodology presented here could be used to analyze other environmental problems. It should also be of interest to the climate change community as it highlights economic constraints and behavior which are crucial to the global warming problem. Finally, we hope that policy analysts working on the global warming problem would use the optimal growth model to provide results which could be used for policy analysis purposes.

This book is the result from the author's doctoral dissertation study at the University of Texas at Austin. I should like thank David Kendrick without whose guidance and support this book would not have been possible. My gratitude also goes to Hans Amman for his suggestions on the modeling aspects of the study, Jurgen Schmandt for identifying the important policy issues, and to Joe Ledbetter for his help in shortening the learning curve on the climate science required for the study. Finally, my wife Chinnie for the moral support and encouragement.

INTRODUCTION

Since the birth of the industrial revolution the atmospheric concentration of a number of greenhouse gases have been rising at a rate unprecedented in earth's history. Scientists have predicted, as early as 1896, that the increase in the atmospheric concentration of these greenhouse gases increases the heat retention property of the atmosphere which inadvertently leads to increases in global surface temperatures. The rise in global temperatures by themselves are not expected to have significant effects on humanity. However, the associated climate changes which accompany the rise in temperatures are expected to cause significant negative effects on the ecosystem as well as the economic system.

Global warming has emerged as one of the most important scientific, political, and economic issue in the last few years. The issue gained importance at an international conference held in 1988 hosted by the Canadian government. The consensus reached at this meeting was that the effects of global warming were second only to a global nuclear war and humanity should take immediate steps towards the elimination of emissions of greenhouse gases. This prompted the United Nations General Assembly to set up the Intergovernmental Panel on Climate Change (IPCC) to study the problem and report back to the General Assembly in 1990.

In August of 1990, Working Group One of the IPCC which comprised of about 300 scientists confirmed the existence of the greenhouse effect and urged policy makers around the globe to adopt policies which will eliminate the emissions of greenhouse gases. The degree of future warming as predicted by the scientists lay in the range of 1.5 degree to 4.5 degrees. This unprecedented rate of temperature increase in the history of the earth has many implications for society. The question that one would ask next is: What caused for the rapid build up of greenhouse gases in the atmosphere?

The positive correlation between economic growth and standard of living had prompted many of the present generation of developed countries to pursue a policy of accelerated economic development through the process of industrialization. One of the unique features of modern industrialization has been its demand for large amounts of energy. This demand for energy has been primarily met by the burning of fossil fuels which incidentally emit large amounts of the greenhouse gas, carbon dioxide; consequently the enhanced greenhouse effect. However, a recent surge in public opinion over the environmental deterioration caused by these policies of accelerated growth has forced policy makers in these countries to reassess the economic growth policy paradigm. Nevertheless, policy makers contend that environmentally sensitive policies are costly and impose huge financial strains on the economy and argue that environment goals can only be achieved at the expense of economic development.

However, the argument, environment at the expense of economic prosperity has been disputed by environmentalists as well as the new generation of ecological economists. The environment versus economic growth argument put forward by policy makers may have been valid in the past when the natural ecosystem was sufficiently large enough to sustain environmental deterioration and also act as a cleaning agent for polluting products: but as the ecosystem is systematically destroyed by the negative externalities of economic activities, it loses both it's cleaning power as well as its natural property to absorb the pollutants. Ecological economists contend that the above argument by policy makers is only valid for short run economic planning. If planning is done over a span of 50 years, the environmental problems caused by economic activities will in fact cause disruptions to the economic system which will in fact in the long run make the costs of achieving economic growth much higher.

This brings us to the issue of the developing countries and their role in the global warming issue. The primary objective of policy makers in these countries is to raise the standard of living of their respective societies. Therefore, noting the success of the developed countries in increasing their standard of living through the process of rapid industrialization, the developing countries are following religiously in the footsteps of their industrialized brothers. This implies that if the same process is followed by the developing countries, the future emissions of greenhouse gases can be expected to increase substantially. This in turn would imply a larger enhanced greenhouse effect and subsequently a more pronounced climate change in the future. This brings us to the objective of this study.

Most of the research work on the global warming problem has been to date conducted primarily by climatologists. Their main focus has been in estimating the degree of warming and the time period in which the warming will be realized in the form of increased global surface temperatures. Climatologists begin their estimation exercise of future temperature increases by using a forecast of future energy demands. Then based on these forecasts, they determine the type and amount of fossil fuels which will be required to satisfy the projected energy demands. Once the type and amount of fossil fuel use has been determined, then the amount of greenhouse gases (carbon dioxide in this case) can be computed from the physical properties of the respective fuels. General Circulation Climate Models (GCMs) are finally used to compute the expected change in the climate systems around the world.

The economists on the other hand have primarily focused on the costs of carbon dioxide (CO2) emission reductions. They use a combination of policy tools like carbon and energy taxes to reduce emissions to specified levels and observe the effects of these taxes on economic performance. In other words, if the climate scientists say that we need a 20 percent reduction in CO2 emissions, then economists specify this reduction level exogenously in their models and determine the taxes which are required to achieve these reductions. With this cost focused methodology, they are also able to determine the economic loss or gain that occurs under the tax regime. However, a criticism of this methodology is that it ignores the benefits from emission reductions which accrue to the economy over time. This is a crucial drawback as it ignores the dynamics of the global warming problem because the costs are incurred immediately while the benefits are observed in the future. Therefore, the argument put forward earlier in the section stating that if natural system is being deteriorated to an extent that it would make economic growth more expensive in the future is ignored in the cost methodology.

Therefore, the question that should be addressed is: what is the optimal level of greenhouse gases emissions that does not perturb the climate as well as the economic system? This book attempts to find an answer to the question above by trying to determine the "optimal" level of carbon dioxide (CO2) emissions which is sustainable by the natural system while simultaneously maintaining specified economic growth objectives. In other words, we focus on an analysis of trade-offs between economic performance and climate change. We do this by formulating a long term growth model as the analytical tool. However, unlike existing growth models which only study the economic system, the growth model in this study consists of a system comprising of an economic and climate sub-system. The two systems are linked by a series of equations which capture the cause-effect relationships between these two systems. The model is called a holistic growth model. The term holistic was put forward by Leopold (1955) to describe a system in which all variables are endogenous to the system even if they belong to other systems. The present style of economic modeling only captures interactions between economic variables.

All economic problems are one of scarcity and choice; thus an appropriate solution procedure which will capture this behavior are optimization techniques. Furthermore, the global warming-economic development discussed in this study is very sensitive to time and therefore it is crucial for the dynamics of the problem be captured. This is uniquely done by formulating the growth model as an optimal control problem and solved using nonlinear programming techniques.

The dynamic growth model in this study is a global economic model, with the globe divided into two regions; developed and developing. Within each region, there are three sectors, agriculture, industry and service. By using an inter-sectoral style of model, we are able to capture the effects of climate change in one sector on the other sectors within the economy. We would have liked to represent the economy of each region with a larger number of sectors but computational constraints forced us to choose this resolution level. Within each sector there are three processes. The pollution intensive process is driven by coal generated energy supply. The pollution intermediate process gets its energy from oil and natural gas. The final process, pollution abatement is assumed to be driven by renewable energy sources.

The main contribution of this book comes from the methodology used in the formulation of the feedback loops in the model which determine the optimal or sustainable level of CO2 emissions. These equations capture the effect of temperature rises and atmospheric carbon dioxide concentration level changes on economic productivity. The presence of a feedback loop between the climate and economic systems ensures that the effects of any climate change on the economic system are captured endogenously within the system. Therefore, in the solution process, an optimal level of CO2 emissions which: (1) minimizes the negative feedback effects of a climate change on the economic system; and (2) minimizes the economic cost of attaining the optimal level of emissions is identified.

The model formulation discussed above will enable policy makers to address the following key issues:

(1) the optimal level of global CO2 emissions which minimizes the effects of a climate change on the economic system;

(2) the issue of partitioning the global CO_2 emission level between the developed and developing regions.

(3) the level of emission reduction aid that is required from the developed to the developing countries.

(4) the optimal technology combination which will ensure the optimal level of CO_2 emission.

(5) allocation of future research funds to study and understand crucial relationships highlighted by the holistic model but which presently have large degree of uncertainty.

(6) to demonstrate that the present style of economic analysis results in inefficient solutions as compared to the holistic style.

This book is comprised of seven chapters. Chapter One gives a brief overview of the scientific theory underlying the greenhouse effect. The empirical evidence suggesting the presence of a strong correlation between economic activities and the greenhouse effect is also presented in this chapter. The chapter ends with a discussion on some of the possible impacts of climate change on the economic system. Chapter Two highlights the important players in the global warming issue. The role of developed countries in the past and their expected role in the future is compared and contrasted with the role of the developing countries in the past and their potential role in the future.

Chapter Three gives the mathematical description of the holistic optimal growth model. The various components of the economic and climate sub-systems as well as the feedback links are discussed in detail in this chapter. A detailed description of each equation is given in this chapter. We also give a general overview of the translation process used in transforming the model from a mathematical format to the General Algebraic Modeling System (GAMS) computer language version in this chapter. Chapter Four can be considered as the data chapter. This chapter describes in detail how the various parameters in the model were computed. We also discuss in this chapter some of the problems and difficulties encountered in data formulation and in the parameter computation process.

Results from a number of policy experiments we conducted are presented in chapter Five. The purpose of these experiments are: (1) to establish the validity of the model; (2) gain insights to the model structure; and (3) compare and contrast a number of different policy options. We conducted a series of sensitivity tests on some of the crucial parameters in the model. Results from the sensitivity analysis are presented in chapter Six. The Final chapter will present a summary of the results and concluding remarks on the global warming-economic development issue.

Chapter One

The Global Warming Problem

1.1 Introduction

Scientists postulate that human activities have been altering the composition of the atmosphere; increased emissions of carbon dioxide by fossil fuel combustion enhance the natural greenhouse effect of the atmosphere and inadvertently raises global surface temperatures. If global warming occurs, it is expected to have significant impacts on society. This chapter's primary aim is first to explain the scientific theory underlying the global warming effect and then go on to discuss some of the impacts global warming will have on society.

The first section of this chapter addresses these questions by beginning with a brief overview of the scientific theory underlying the greenhouse effect. The scientific theory is then used to explain how emission of greenhouse gases by economic activities cause the anthropogenic (man-made) greenhouse effect[1]. Once the relationship between economic activities and global warming has been established, historical data is used to substantiate the theory. Due to the complex nature of the scientific theory of the enhanced greenhouse effect as well as the relationships between economic activities and global warming, there is a high degree of uncertainty involved in estimating the magnitude and timing of the anthropogenic greenhouse effect. Therefore, after establishing the relationship between economic activities and global warming, some of the inherent uncertainties in estimating the magnitude and timing of global warming are discussed.

The second section then focuses on the various impacts of global warming and the effect on the economic system. The section also discusses some of the difficulties encountered when estimating the magnitude of these impacts on

[1]The term anthropogenic greenhouse effect is used interchangeably with global warming in this study.

society. We shall discuss the source of these uncertainties and the methodology used in this study to mitigate the degree of uncertainty in this section.

1.2 The Greenhouse Effect

The theory of the greenhouse effect was first introduced by scientists at the turn of the century (Arrenhius 1896). Rather than giving a detailed scientific explanation of the mechanics of the greenhouse effect which is beyond the scope of this book, a simplified version of the process is given below.

The atmosphere is comprised of a number of gases, some of which act as an insulation blanket over the Earth's surface. These gases have been present in the Earth's atmosphere in trace amounts for the better part of the Earth's history. The major trace gases are water-vapor, carbon dioxide, methane, nitrous oxide, and chlorofluorocarbons.

A subset of the atmospheric gases called greenhouse gases (water-vapor, carbon dioxide, methane, nitrous oxide, chlorofluorocarbons) allow the sun's ultra-violet and visible radiation to enter the Earth's atmosphere but retain infra-red radiation emitted by the Earth's surface. Let us illustrate the mechanics underlying the radiation budget balance are best illustrated by a simple example. The numbers used in the example are not the actual amounts but numbers chosen to illustrate the equilibrium balance that governs the Earth's radiation budget. Figure 1.1 is a schematic diagram of the Earth's radiation budget[2].

---

[2]The numbers I use in this example are similar to U.N.E.P's example in the publication *GreenhouseGases* 1987.

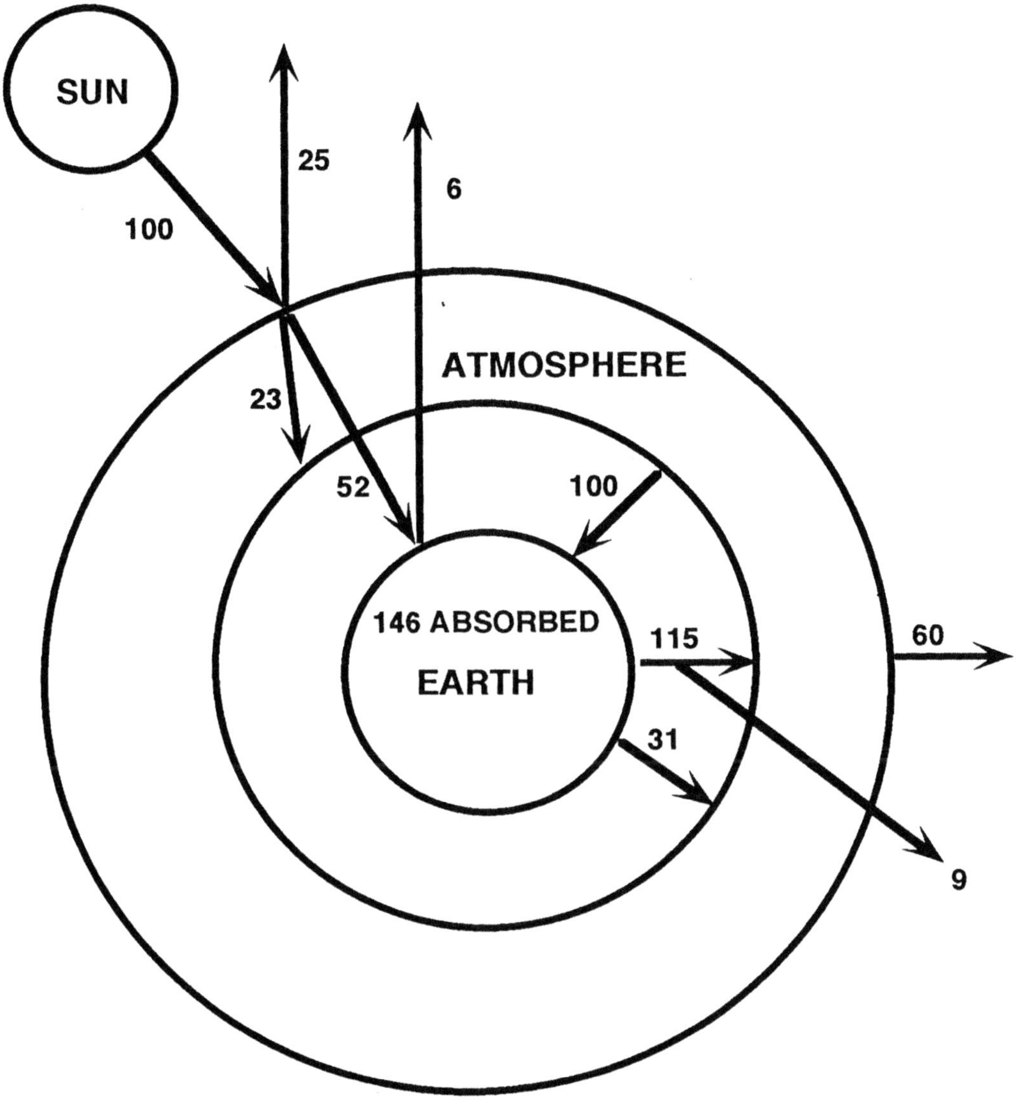

Figure 1.1 The earth's radiation budget balance.

One hundred units of radiation from the Sun strike the atmosphere. Of the 100, 25 units are reflected back to space. This level is dictated by the albedo effect of clouds and particles in the atmosphere. A further 23 units are used in heating up the atmosphere. Of the 52 units left, 6 units are reflected back into space due to the albedo effect of the earth's surface. This leaves a total of 46 units to warm up the earth's surface.

There is at the same time a dynamic relationship occurring between the atmosphere and the Earth's surface. Both the atmosphere and the surface of the Earth warm up and radiate infra-red energy at each other. The earth radiates 115 units of infra-red energy to the atmosphere. The atmosphere absorbs 106 of these units and the remaining 9 escape out to space. Of the 106 units, 60 units are radiated by the atmosphere out into space. The remaining 46 units are absorbed by the greenhouse gases in the atmosphere; this is the cause of the greenhouse phenomena. The first law of thermodynamics is observed at all boundaries. This is shown below:

1) If the earth absorbs 146 units, then to maintain equilibrium it must also radiate 146 units.

1a) The 146 units the earth absorbs is comprised of the 100 units from the atmosphere and 46 units from the original incoming solar radiation.

1b) The 146 units the earth radiates is made up of the 115 units towards the atmosphere and 31 units by convective processes on earth

2) The atmosphere then absorbs 160 units and to maintain equilibrium, it radiates 160.

2a) Of the 160 units the atmosphere absorbs, 106 units are from earth, 23 units from the original incoming solar radiation, and 31 units from the convective processes of the earth's surface.

2b) The 160 units the atmosphere radiates is comprised of 100 units towards earth and 60 units towards the sun.

3) Based on the above figures, if the sun radiates 100 units towards earth and its atmosphere, than earth and its atmosphere must radiate 100 units to maintain equilibrium.

3a) The 100 units radiated to space is comprised of 60 units from the atmosphere, 9 units from earth's 115 units it radiates towards the atmosphere, 6 units directly reflected from earth's surface, and finally the 25 units reflected by the atmosphere.

Water-vapor, the most abundant of the gases, is the most important natural greenhouse gas. Carbon dioxide is the second largest and is added to the atmosphere both naturally and by human activities. Methane is the third largest component followed by nitrous oxide and chlorofluorocarbons respectively. Without these gases, the earth's surface would be too cold to support life as we know it. Therefore, the greenhouse effect is essential to life.

However, the atmospheric concentrations of all these trace gases, except water-vapor have been rising in the last hundred years. The primary reason for the buildup of these trace gases has been the increase in emission of these gases by human activities. The increase in these trace gas concentrations in the atmosphere can affect the amount of heat which is trapped by the atmosphere. The example above illustrates a stable system which is in equilibrium. The following example illustrated in Figure 1.2 shows how an increase in atmospheric carbon dioxide caused by economic activities destabilizes the system and causes global warming.

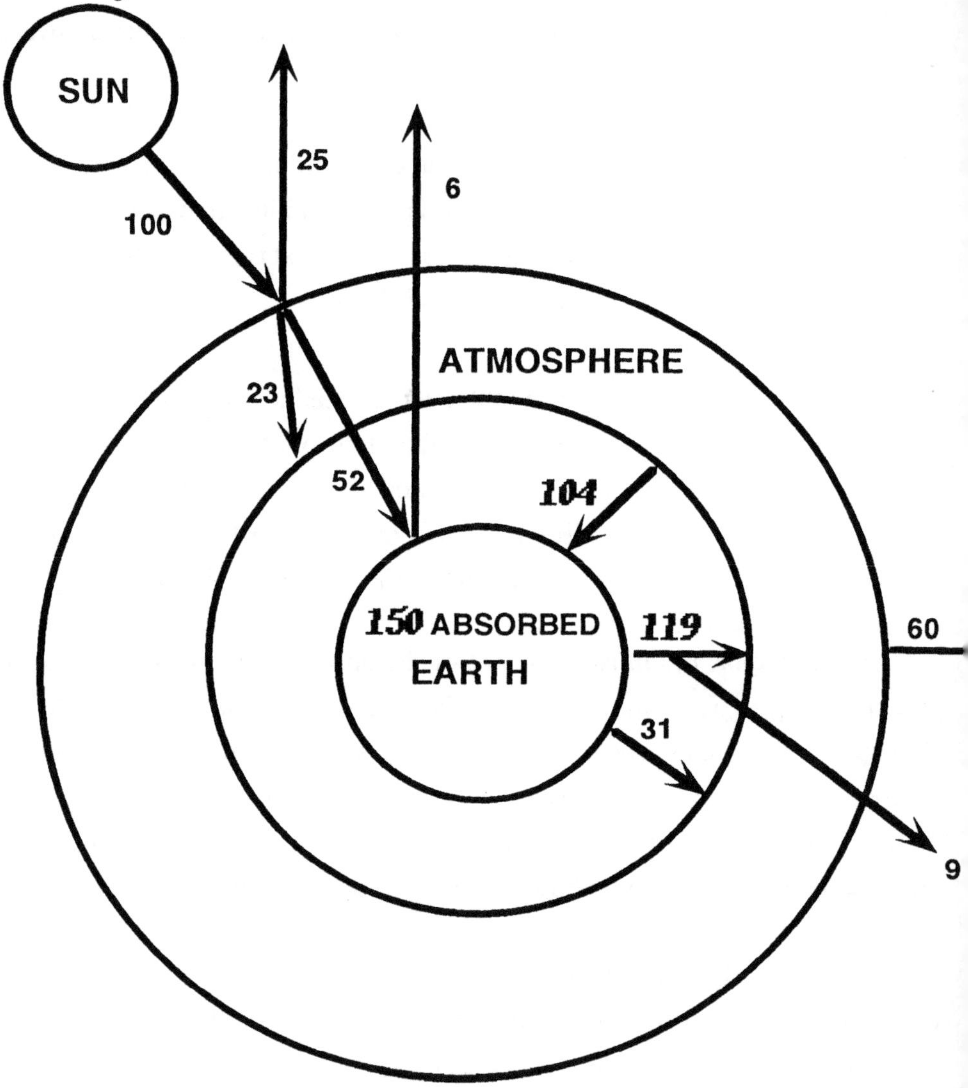

Figure 1.2 Radiation budget after perturbation.

1) If earth radiates as before 115 units towards the atmosphere, then due to the enhanced greenhouse

effect, 50 units are absorbed by the atmosphere rather than 46 units[3]. The atmosphere then only radiates 56 units out to space and not 60 units. The first law of thermodynamics is violated. A disequilibrium occurs.

2) The atmosphere reacts to the disequilibrium by radiating the extra 4 units to earth. Therefore, the atmosphere now radiates 104 units toward earth and not 100 units as it did before the disequilibrium.

3) If the earth now absorbs 150 units, then to maintain equilibrium it radiates 150 units .

3a) Of the 150 units earth absorbs, 46 units are from the sun's incoming solar radiation and the remaining 104 units come from the atmosphere.

3b) Of the 150 units earth radiates, 31 units are from convective processes. This amount remains unchanged. However, the infra-red radiated by the earth's surface increases by 4 units, the same amount that is absorbed by the increased concentration of greenhouse gases. This is the mechanism by which surface temperatures increase.

4) The atmosphere now absorbs 164 units. 31 units from convective processes, 23 units from the sun's initial solar radiation, and 110 (9 units are still

---

[3]The increase in 4 units is arbitrary and is chosen for illustrative purposes.

radiated towards space) units from infra-red emission.

4a) The atmosphere is now back in equilibrium. It radiates 104 units to earth and back to the original 60 units to space.

5) Total radiation towards space is now back to 100 units. The initial deficiency of 4 units caused by the increased absorption of the greenhouse gases has been corrected.

It can be inferred from the simple example above that an increase in any of the greenhouse gases decreases the long wave flux (F) from the atmosphere out into space. Thus an increase in atmospheric concentration of these gases decreases the energy loss from the troposphere to space. To maintain equilibrium, dictated by the first law of thermodynamics, the earth's surface heats up by the same amount (F). The increase in F caused by higher temperature balances the decease in F.

1.3 Historical Data Substantiating the Greenhouse Effect

Empirical evidence supporting the scientific theory of the greenhouse effect can be divided into two major categories. The first category involves the collection of data over a time span of a thousand years. Scientist have been able to collect this data from air trapped in ice-core samples taken from glaciers. Preliminary results from these samples suggest a high degree of correlation between past atmospheric carbon dioxide concentration levels and temperature levels. Warm periods during earth's history were accompanied with high levels of atmospheric carbon dioxide. These high levels of carbon dioxide have been primarily attributed to high volcanic activity in the past. For a more detailed analysis of these experiments, the reader is referred to studies by Oeschger (1984) and Hammer (1980).

The next category deals with empirical results collected over a shorter time period, a time span of approximately a 100 years. Accurate measurements of atmospheric carbon dioxide concentrations were first collected by Keeling on Mauna Loa in 1957. Figure 1.3 below shows the trend observed on Mauna Loa.

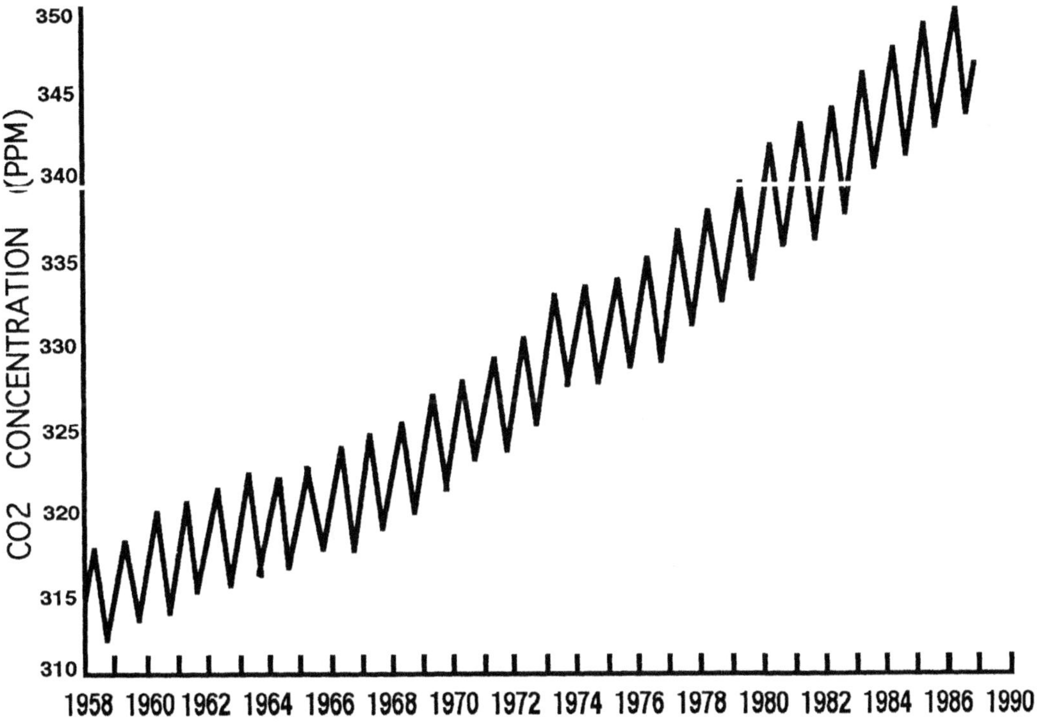

Figure 1.3 Atmospheric CO2 Concentration Trend Observed at Mauna Loa, Hawaii

There were earlier measurements of carbon dioxide concentrations but these were not accurate and the methods used to measure them were not precise. However, based on Keelings historical trends, Wigley (1983) was able to approximate the carbon dioxide concentration in 1870 as 270 ppmv. With respect to temperature data, Hansen and Lebedeff(1987) were able to create a time series of monthly global surface temperatures from 1880 to 1980. Figure 1.4 shows the results derived by the group at the climate research unit at East Anglia University which reflect an upward trend in temperature in the last 100 years. We can infer from the two graphs that there is a high degree of correlation between temperature and atmospheric carbon dioxide concentration.

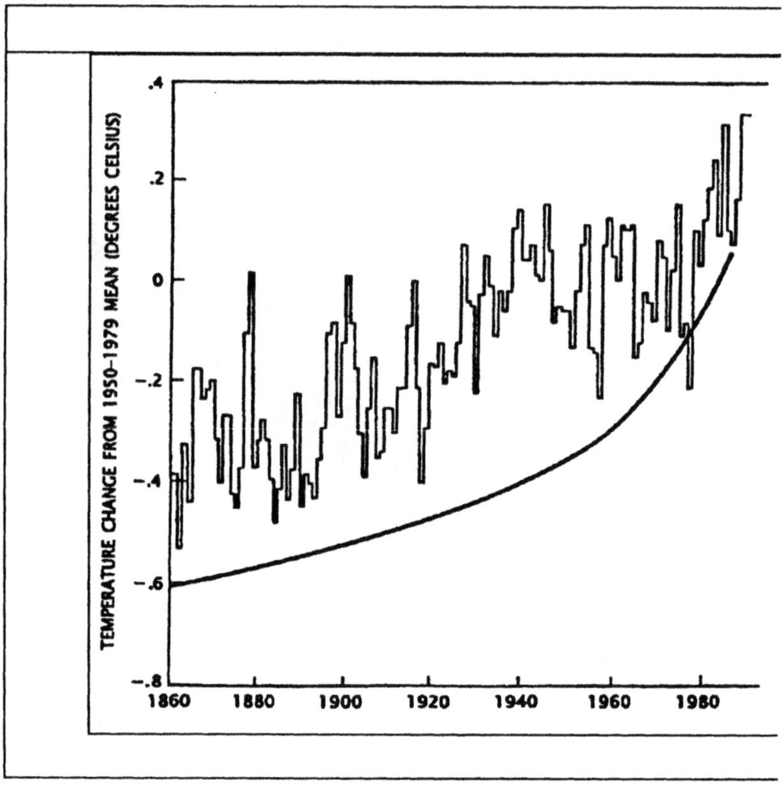

Figure 1.4 is reprinted from the greenpeace report on global warming. The increase from 1990 to the year 2000 in the graph is based on estimated fossil fuel from past trends.

Figure 1.4 Source: Global Warming: The Greenpeace Report 1990.

What caused the increase in atmospheric carbon dioxide concentration in the last 100 years? Figure 1.5 shows that the time the interglacial level of atmospheric carbon dioxide concentration started to rise coincides with the start of the industrial revolution. Since then, the atmospheric carbon dioxide concentration has increased by approximately 25 percent, from 270 ppmv to 350 ppmv.

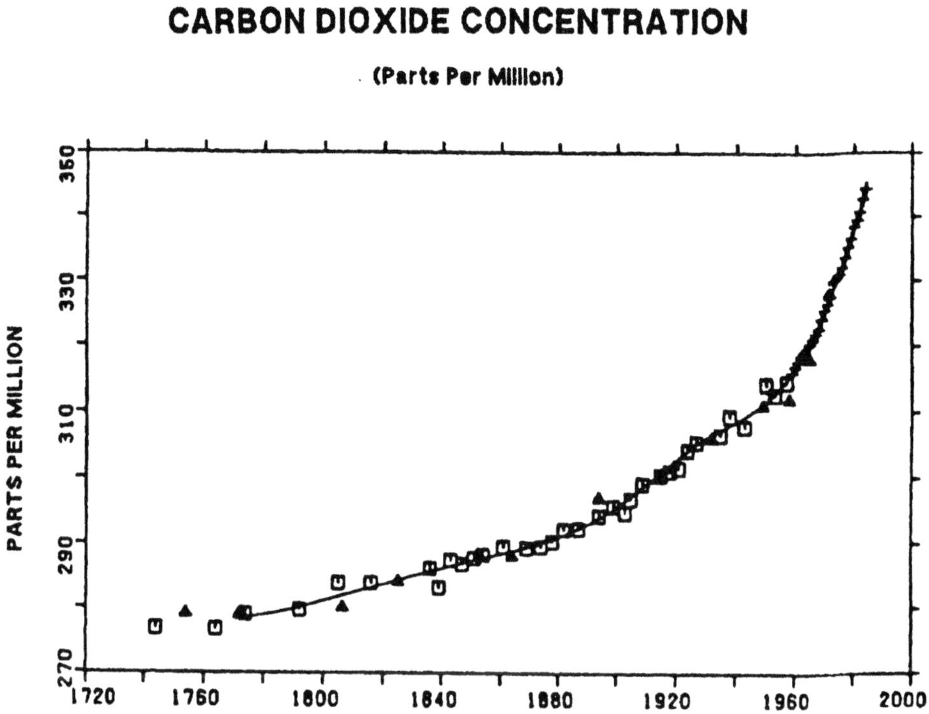

Figure 1.5 Source: Lashof and Tirpak 1990 pg 43

The primary source of carbon dioxide emissions during the period illustrated in Figure 1.5 was from the burning of fossil fuels to satisfy energy demand. Modern economic growth synonymous with industrial growth, demands huge amounts of energy. The earliest source of energy was coal which is also the largest emitter of carbon dioxide. With the introduction of new technology, oil and gas have replaced coal as the major source of energy but even if these new energy sources emit less carbon dioxide than coal they still emit sufficiently large amounts of CO2 to cause global warming. Thus, unless economic growth can be accomplished without imposing the usual huge energy requirements, policymakers will need to either introduce CO2 emission free energy sources or reduce the present growth trend in fossil fuel use.

Based on the discussion so far, one would logical conclude that we have to take steps to reduce greenhouse gas emissions in particular CO2. However, although is a general consensus among the scientific community that a global warming will occur if present trends in fossil fuel use continue, there is considerable disagreement over the issue of the time period when a warming will occur and the magnitude of the warming. This disagreement would not pose a problem for reducing emissions if the costs of reduction are minimal. Unfortunately, preliminary studies suggest that abatement costs could be considerable and may have significant effects on economic performance. Following these results, a number of countries have suggested that some of the inherent uncertainties in the enhanced greenhouse effect be resolved before any emission reduction policies are enforced. As these uncertainties play a big role in policy making, we thought that it would be useful to discuss these uncertainties in some detail.

1.4 Uncertainties in Estimating Global Warming

The primary source of uncertainties is the lack of knowledge over the dynamics involved in the economic, carbon cycle, and climate systems. The degree of warming and the time the warming occurs depend largely on the uncertainties present in the dynamics of the three systems mentioned above. We shall begin by highlighting some of the uncertainties inherent in the economic system, followed by the problems faced in the carbon cycle and climate system. We shall also discuss some of the methodologies used to mitigate the degree of uncertainty in these systems.

1.4.1 Uncertainty in the Economic System

The major source of uncertainty in the economic system lies in the difficulty in forecasting future energy demands over a time span of 50 years. Future energy demands are based on a complex set of variables which are unknown and difficult to project from past trends[4]. Furthermore, once the level of future energy demand has been estimated, the next step is to estimate how much of the projected energy demand will be met by fossil fuels, because it is the use of fossil fuels which causes the anthropogenic greenhouse effect.

In light of the difficulty in estimating future energy demands, present generation modelers use the next best alternative. They identify upper and lower bounds for future energy demands. Based on these upper and lower bounds, carbon dioxide emission levels are then calculated[5]. This methodology therefore determines the magnitude of global warming by estimating future fossil fuel use. Figure 1.6 gives a schematic illustration of the process.

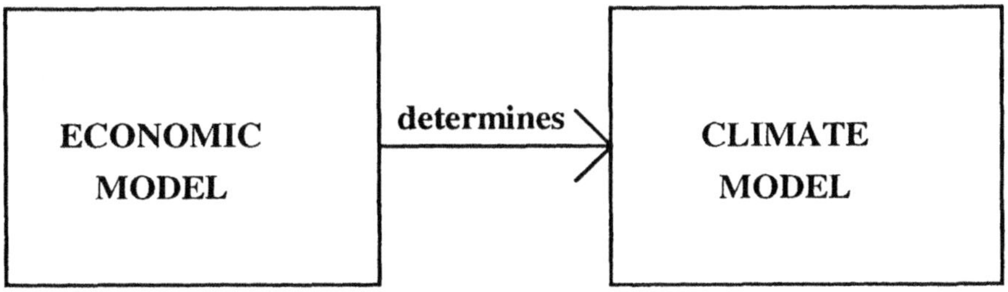

Figure 1.6 A Traditional Economic Model Structure

This study, however, approaches the problem from a different perspective. Rather than attempting to forecast energy demands and thus fossil fuel use, the model in the study attempts to identify the optimal level of global energy use and

[4]Some of the variables which affect projected future energy demands are population growth, technological advance, potential fossil fuel resources and the price of fossil fuels.
[5]This procedure was used extensively in studies conducted by 1)Nordhaus and Yohe and 2) Edmonds and Reilley. Refer to the bibliography for a full reference.

the appropriate fuel combination[6] which is compatible with a stable climate and economic system. The process used in this study is illustrated in Figure 1.7.

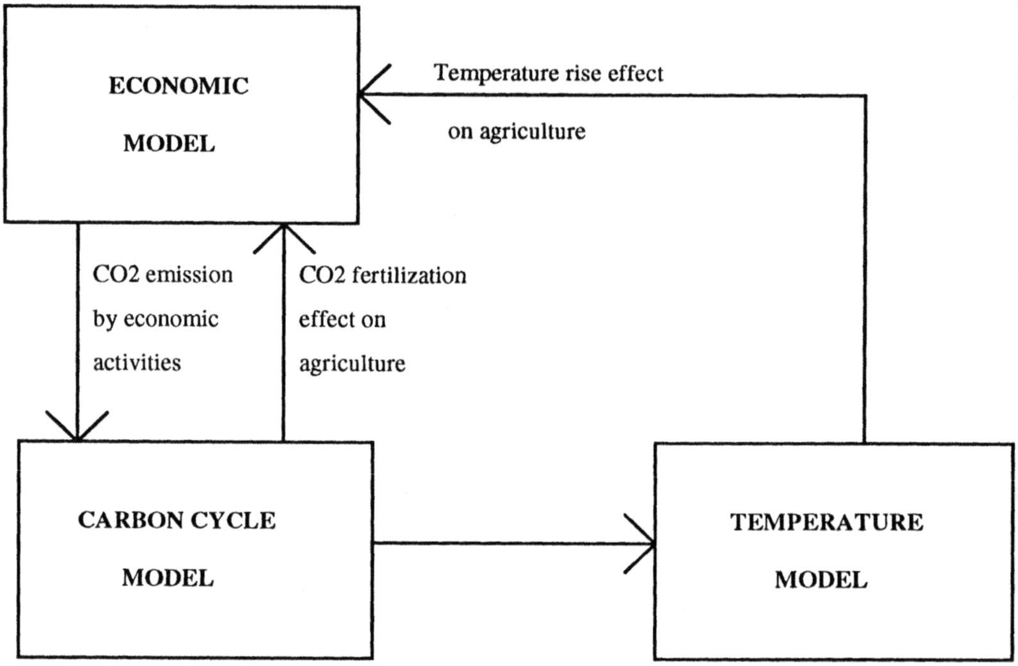

Figure 1.7 The Holistic Model Structure

By using the proposed methodology, we are able to project future energy use which is sustainable by the natural system. The uncertainty inherent in the first methodology is mitigated to a certain extent. However, because the sustainable level is dependent on the carbon cycle and temperature models, the uncertainty in these two systems will play a role in determining the sustainable level. This now leads us to the carbon cycle model.

---

[6]In the study of the energy sector, three different types of fuels are considered. The scenarios are classified by the pollution type processes. The pollution intensive in the energy sector uses coal, pollution intermediate uses oil and natural gas. and pollution abatement uses solar and hydro sources.

1.4.2 Uncertainty in the Carbon Cycle System

Once the emission level of carbon dioxide by economic activities has been established, the next question is: how fast will the atmospheric carbon dioxide concentration increase? Not all carbon dioxide emitted by economic activities remains in the atmosphere. There is a dynamic process of CO2 transfer among the atmosphere, land, and oceans.

The natural sources of carbon dioxide are the oceans and land plants. The unnatural source is the economic activity of society. The natural sinks are the oceans, land plants, and the atmosphere. This dynamic exchange process constitutes a carbon cycle system and is an integral part of the global warming problem. The carbon cycle system ultimately determines the amount of the carbon dioxide emitted by economic activities which remains airborne. The uncertainty which arises from the carbon cycle system is due to the lack of knowledge by scientists of the role oceans and land play in the absorption and emission of carbon dioxide.

The primary source of uncertainty arises from the limited knowledge on the part of scientists of the physical, biological, and chemical processes which govern the transfer rates of carbon dioxide among the reservoirs mentioned above. Although scientists have made considerable progress in building more sophisticated carbon cycle models in the last decade, these models have yet a far way to go before they can predict future atmospheric CO2 concentrations with a high level of accuracy. However, climatologists conclude that the degree of uncertainty will not result in significant prediction errors over a fifty year planning period (Edmonds 1986 pg 104).[7]

---

[7]For a detailed overview of carbon cycle models , the reader can refer to Bolin (1981).

The secondary source of uncertainty arises from the introduction of feedback processes. These feedback processes fall into two groups, positive and negative feedback effects. For example, increased atmospheric carbon dioxide will increase the size of plants as CO_2 is a plant fertilizer in the photosynthesis process. This increase in plant size will increase intake of carbon dioxide and will to a certain extent decrease the buildup. This is an example of a negative feedback. On the other hand, a carbon dioxide build-up can cause a warmer climate which in turn can cause disruptions to forest growth thus reducing the carbon dioxide intake and therefore increase the carbon dioxide buildup. This is an example of a positive feedback. These uncertainties are hard to quantify and due to lack of data are difficult to estimate. Scientists have therefore ignored these feedback effects in their carbon cycle models and have assumed constant exchange rates among the reservoirs.

Although estimating atmospheric carbon dioxide concentration is still not an exact science, the present generation of carbon cycle models give approximate results which are helpful in estimating temperature rises.

1.4.3 Uncertainty in the Climate System

In the earth's radiation budget example above, there are a number of factors which can change as temperature rises. For example, the albedo effect of the earth's surface. As temperatures rise, the ice caps at the Arctic will melt thus reducing the amount of ice. A reduction in ice reduces the amount of incoming solar radiation which is reflected back into space. This is an example of a positive feedback effect of an initial temperature rise.

Another factor which is of concern among scientists is the role of clouds. As temperature rises, there is increased evaporation. The increased evaporation rate forms more clouds. Increased cloud cover has two effects. The first effect is positive and is caused by the insulatina property of clouds. Clouds trap outgoing radiation and therefore, with more clouds, more outgoing radiation is trapped. The second effect is negative and is caused by the albedo property of clouds. Clouds reflect incoming solar radiation and therefore, with a larger cloud cover, a larger amount of incoming solar radiation is reflected back into space. The overall effect of increased cloud cover is still under investigation but Ramanathan (1989) concludes that the negative effect of increased cloud cover is slightly more pronounced than its positive effect.

The third major source of uncertainty in the climate system is the role oceans play. Oceans are known to moderate the global warming effect to a certain extent and are believed to slow down the initial scale of warming. Scientists are still investigating the heat conducting process between the surface and the deep ocean. Most climate models simulate ocean dynamics by assuming: (1) no heat flow between the surface and the deep oceans; (2) an ocean depth of only about 70 meters; and (3) no horizontal transport of heat between the latitudes. Scientists believe that as oceans are an extremely important component in climate change, ocean heat circulation dynamics should be modeled separately and then linked to a climate model.

This concludes the first part of the chapter. The uncertainties discussed above do not alter the view that global warming is imminent. They do however, give rise to disputes over the degree and timing of the warming. These disputes can be minimized by carrying out sensitivity analysis on the parameters which have a high degree of uncertainty. By using upper and lower bounds on these parameters, economists can derive plausible ranges for future climate and economic scenarios.

1.5 Impacts of Global Warming

Is the anthropogenic greenhouse effect a problem for society? And if it is, what are the impacts and the magnitude of these impacts? In spite of the imperative for economists and policymakers to have knowledge of the consequences of a climate change to society. This section will address the questions stated above. The discussion will however concentrate on the effects global warming will have on economic activity, primarily the agricultural sector.

The study will also focus on the capabilities of regions to respond to a temperature rise and the costs incurred. The main emphasis is on the developing countries. Most of the developing countries lie in marginal agricultural areas which with slight disturbances in the climate system can cause drastic drops in crop yields. However, the reader should note that there is a high degree of uncertainty involved when estimating the magnitude of the impacts of a climate change on the agricultural sector. Some of these uncertainties are discussed in this section.

Scientists have estimated some of the consequences of a climate change by using climate impact studies. Climate impact studies present in a systematic manner the consequences a climate change caused by human activity can have on society and the natural ecosystem. Two methodologies are used presently to compile climate impact studies. The first method involves using historical climate data to postulate future impacts of a carbon dioxide induced climate change. The second method uses climate simulation models to generate possible future climate impacts.

The present generation of climate models are able to predict climate changes and the corresponding impacts on society and the natural ecosystems to areas as small as the state of Colorado. But due to the high degree of uncertainty present in estimating the degree of temperature rise, most models use scenario analysis to derive climate impact statements. A scenario analysis entails setting a plausible range over which the unknown parameter can vary and then estimating the impacts within this set range.

Three major impacts have been identified which may cause substantial damage to both ecosystems as well as the socioeconomic system of man. The first impact is the effect of climate change on agriculture. The second impact is on the ecosystem and the third impact is on sea-levels. This study will concentrate on the impact global warming will have on agriculture which has significant implications to the socio-economic systems.

1.5.1 Agricultural Output (Food Supply)

Climate plays an important part in determining agricultural output. Therefore, any climate change will have significant impacts on agricultural yields. There are two approaches to analyzing the effects of increased atmospheric carbon dioxide concentration levels and increased temperatures. The first approach is called the "slow change view" and the second is called the "shift in risk view" (Bolin 1986).

1.5.1.1 Slow Change View

Proponents of the slow change view argue that it is the direct effect of the change in atmospheric carbon dioxide concentration and the corresponding temperature rise which affect crop yields. The slow change view recognizes that climate change induced by increased atmospheric carbon dioxide concentration introduces two feedback effects which affect crop output. The first effect is called the carbon dioxide fertilization effect. The second effect is caused by the effects of increased temperatures.

1.5.1.1.1 Carbon Dioxide Fertilization

Carbon dioxide acts as a natural fertilizer for crops and in a carbon dioxide enriched environment, crop yields increase. However, the impact varies across crops with C_3 crops increasing yields by approximately 10 to 30 percent and C_4

crops by 10 percent. In this study, emphasis will be placed on food crops, which are primarily C_3 crops, because: (1) these crops constitute the bulk of world food production; and (2) as these crops are the major food supplies for many nations, any fluctuations in the yields of these crops will have significant economic and political effects on society.

In greenhouse experiments, carbon dioxide fertilization has increased yields of wheat by 30 percent, and yields of rice and maize by approximately 10 percent (UNEP 1990). Agronomists however are cautious in interpreting these results as these studies were conducted in controlled environments (greenhouses) with no fluctuations in other crop yield sensitive parameters like soil moisture and precipitation. Furthermore, the carbon dioxide fertilization effect exhibits diminishing returns and inadvertently the fertilization effect will cease to be a major factor.

1.5.1.1.2 Temperature Rise Feedback Effects

The effects of an increase in temperature are primarily negative as compared to the positive effect of the carbon dioxide fertilization effect. Increased temperatures translate to increased soil moisture evaporation and the resulting dry soil loses its nutrient qualities rapidly. Furthermore, as surface temperatures rise, present crop growing belts will shift northwards, for example the wheat belt on the American continent. However, the problem with the shift northwards is that the new lands which will be in the appropriate temperature belt, have inappropriate soil conditions for growing crops. To prepare these lands for agricultural use will entail huge investments. Therefore, even if global food supply can be maintained, the cost of production will definitely increase.

Another side effect of a climate change is the change in precipitation patterns. According to General Circulation Models (GCM) most of the marginal agricultural regions will experience a drop in annual precipitation and an increase in evaporation rates which further exacerbate the precipitation problem. The regions which are affected significantly by precipitation changes are regions which do not have the adequate infra-structure to adapt to the decrease in precipitation. Irrigation schemes entail vast amounts of investment and though the developed countries are able to adapt to the new drier environments by installing elaborate irrigation structures, the developing countries will have a more difficult time carrying the burden of the extra expenditure. The recent MINK (Rosenberg 1990) study indicates that the grain growing belt in the USA will experience a drop in grain production of approximately 8 percent even with its elaborate irrigation schemes. In the case of the developing regions the drop in production in the developing regions can be expected to be larger. Numerous studies have been conducted to investigate the effect of precipitation fluctuations on crop yields in the developing region (Parry 1988).

1.5.1.2 Shift in Risk View

Proponents of this approach argue that it is not the direct effects of a climate change which are of consequence but the increased frequency of adverse climate events. The increased occurrences of droughts, floods, and frosts damage are expected to have larger effects on the yields of food crops. Climatologists however have so far been unable to predict the change in frequency of these events as the climate system is perturbed. Past studies indicate that it has been the occurrences of adverse climate events which have caused major food supply disruptions. Therefore, if a climate change caused by the accumulation of greenhouse gases is going to increase the occurrences of these adverse events, then policymakers must factor in the costs of these disruptions when making the decision as to whether the costs of preventing the greenhouse effect outweigh the benefits.

1.6. Conclusion

This chapter has focused on the scientific aspects of the greenhouse effect, its impact on agriculture, and some of the uncertainties surrounding the issue. We can conclude from the discussion on uncertainty that there is a high degree of uncertainty surrounding the global warming problem, However, as Schneider puts it very eloquently, "Modeling is a major advance over hand waving forecasts of global change" (Leggett 1990).

With respect to the problem of forecasting global warming by predicting future level of economic activity, this study attempts to determine the optimal level of emissions by conducting an analysis of tradeoffs between carbon dioxide emission reduction costs with damage costs incurred if a global warming does occur. Therefore, unlike previous studies, this study captures the feedback effects of an increase in atmospheric carbon dioxide concentration and temperature rise on economic activity explicitly within the model structure. And of the two approaches put forward by which one can evaluate the magnitude of global warming on agricultural yields, i.e., the feedback effects, this study uses the slow change view.

The next chapter will discuss some of the policy responses which can be adopted by nations around the globe. The chapter also discusses the difficulties which may arise when an international effort is undertaken to curb global warming.

Chapter Two

Policy Responses and International Coordination

2.1 Introduction

The general consensus in the scientific community over the inevitability and irreversibility of the enhanced greenhouse effect and the subsequent impacts on society make it imperative for policy makers to respond. However, the level and intensity at which they should respond is the key issue. Should they implement policies immediately to curb carbon dioxide emissions or should they adopt a wait and see attitude till all the uncertainties mentioned in Chapter One are resolved? And if they do decide to act now, will regional emission reduction policies be effective in solving the problem? If not, should they try to negotiate an international emission reduction treaty or just implement regional policies which will act as buffers for their respective societies from the impacts of global warming? If the answer is YES, then questions regarding the issue of international cooperation and coordination will have to be addressed and answered. The list of question is endless but these are some of the many questions and issues this chapter will try to address.

We begin by identifying the two major policy strategies which have been put forward by policymakers and scientists to address the global warming issue. Next, the advantages and the disadvantages of these policy responses are discussed. Following this, the policy strategy and the appropriate tools used in this study are highlighted. Once the strategies and the tools have been identified, the question of international cooperation is addressed; following which, additional policy tools to ensure cooperation are highlighted. The main idea underlying the discussion on cooperation is that without a coordinated international effort, the global warming problem cannot be resolved.

2.2 Policy Objectives

The various policies which have been suggested in response to the global warming problem fall into two major groups: 1) adaptive policies; and 2) preventative policies. Adaptive policy measures are primarily designed to help prepare regions to adapt to a climate change. Preventative policies, on the other hand, primarily focus on slowing down and finally eradicating the anthropogenic greenhouse effect.

2.2.1 Adaptive Policies

As mentioned in the above paragraph, policies in this group are primarily designed to provide relief from some of the impacts of global warming. Some examples which fall in this category are:

1) Coastal Area Management. Building dikes to protect coastal lands from temperature induced sea-level rises.

2) Water management. Building irrigation and drainage systems to buffer fluctuations in rainfall caused by global warming induced climate change.

3) Soil Management. Protective measures to prevent loss of soil quality caused by higher temperatures.

From the examples above, it can be inferred that adaptive strategies are short run strategies which prepare regions to adapt to global warming induced adverse climate events. However, they are by themselves not capable of eradicating the global warming problem.

If adaptive policies cannot eradicate global warming, than what purpose do they serve? There are two answers to the above question. The first response is that adaptive costs are significantly less than the preventive costs which are needed to reduce GHG emissions. This is still a highly debatable point depending on the degree of temperature rise. The second response points to the fact that GHG emissions to date have already committed us to a certain degree of global warming. Scientists estimate that the total emissions of greenhouse gases from 1880 to 1989 has committed earth to a 0.5 to 1.5 degree rise in the future. They estimate that the present realized increase of 0.3 to 0.5 degrees over pre-industrial periods is due to the increase in greenhouse gas emissions between 1880 and 1989 (Legget 1990). Therefore, even if there is a complete stop in greenhouse gas emissions today, the earth is already committed to a 0.5 to 1.5 degree rise in the future. Moreover, if the general consensus of the scientific community is that this temperature rise is inevitable, it is imperative to implement adaptive policies to mitigate the effects of global warming on society.

But critics of the global warming debate argue that adaptive polices burden the economic system with unjustified costs. These critics contend that the two to five degree rise as postulated by climatologists is highly speculative and is subject to debate (ibid). They argue that the high degree of uncertainty associated with the estimated figures for committed temperature rises does not justify the expense incurred by adaptive policies.

On the other hand, proponents of global warming argue that any policy measure which buffers society from the impacts of climatic fluctuations is beneficial with or without global warming. They contend that adaptive policy measures which buffer society from the impacts of global warming will also buffer society from present adverse climate events like droughts, floods, coastal erosion, and soil erosion. These events which occur even in the absence of global warming cause huge losses and disruptions to economic systems. Therefore, if adaptive policies are implemented, then present occurrences of adverse climate events will have smaller repercussions on society. Although these policies may impose financial burdens on the economic system in the short run, the advantages of attaining an economic system which is less vulnerable to climate fluctuations in the long run far outweigh the short run costs.

2.2.2 Preventative policies

It can be inferred from the above paragraph that adaptive policies only mitigate the effects of global warming on society. However, we should be aware that if only adaptive policies are adopted without any attempt being made to eliminate global warming, there is a critical threshold beyond which the effectiveness of adaptive policies to mitigate the effects of global warming will decrease. The ideal solution calls for a combination of adaptive and preventative policies.

Preventative policies are primarily designed to slow down and finally eliminate the threat of an anthropogenic greenhouse effect. As illustrated in chapter one, the main cause of global warming is the increase in emission of greenhouse gases by economic activities. Thus, to minimize the feedback effects of global warming, greenhouse gas emissions by economic activities must be reduced to sustainable levels. However, this option may cause significant economic growth retardation because the largest emitter of the major greenhouse gas (carbon dioxide) is the energy sector; and the energy sector plays a crucial part in the economic growth process.

The above "paradox of the environment" has been used by critics of global warming as an argument against imposing emission limits on greenhouse gases. These critics contend that the economic loss caused by emission cuts will be much larger than the economic damage caused by the feedback effects of global warming. This line of argument is valid if policymakers are primarily concerned about short run economic growth. However, if the planning period is lengthened to 50 years or more, the damage caused by the feedback effects of global warming may far outweigh any benefits accrued in the interim from economic growth. Thus, if a cost and benefit analysis is to be done for evaluating global warming preventative policies, the time period over which the evaluation is done must be long enough to capture the feedback effects of global warming on the economic system.

We focus on global warming preventative policies which reduce the emission of greenhouse gases to levels which are sustainable by the natural system. The main greenhouse gas which this study will focus on is carbon dioxide. The reason for choosing carbon dioxide is: 1) it is the largest contributor to the anthropogenic greenhouse effect as shown in figure 2.1; 2) the carbon cycle is relatively well understood as compared with the other greenhouse gases; and 3) empirical data on carbon dioxide is relatively concise as compared with the other greenhouse gases for which there is little or no data.

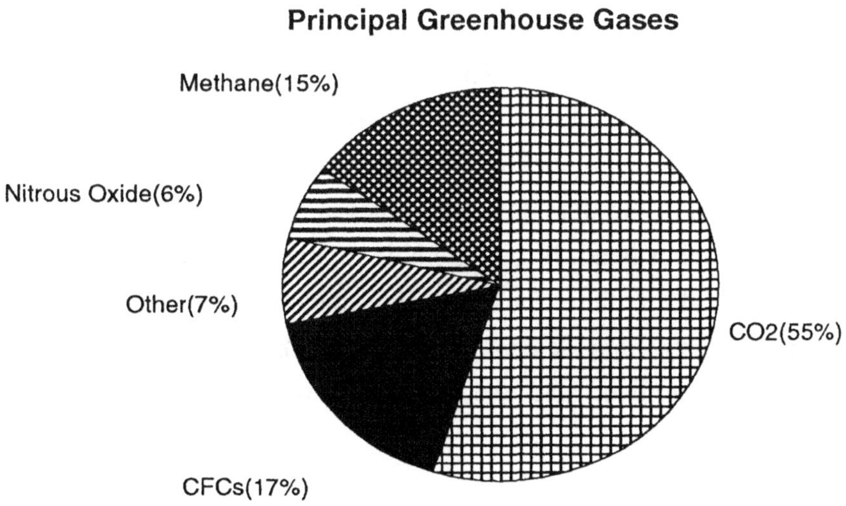

Fig 2.1 Relative contributions by the major greenhouse gases to the anthropogenic greenhouse effect.
Source: Lashof and Tirpak. 1990

Carbon dioxide can be reduced by four methods:

1) Conversion from fossil fuels like coal which emit high levels of carbon dioxide to fossil fuels like natural gas which emit less.

2) Energy conservation and efficiency programs. These programs ensure fuel efficiency.

3) Reforest the planet. This increases the intake of carbon dioxide from the atmosphere.

4) Replace fossil fuels with renewable energy sources to meet future energy demands.

All four methods are designed to reduce the level of carbon dioxide emissions. However, it is the fourth method which is primarily designed to reduce CO2 emissions substantially from present levels. The adoption of these policies and the time period over which they are enforced will determine to a large extent the cost of a global warming eradication policy. If policymakers decide to initiate all four policies immediately, then the costs incurred by these policies may outweigh the damage costs which a global warming may cause in the future. On the other hand, if they are able to introduce the four methods over an "optimal" period of time thus allowing the economic system to switch gradually from "dirty"[1] to "clean" technology, the costs incurred by these methods can be minimized and therefore not interrupt the economic growth process too drastically.

Nevertheless, a policy strategy consisting of all four methods mentioned above will impose some restraint on economic growth initially. Developed

[1]Dirty technology means carbon dioxide intensive emission processes.

countries can to a certain extent be able to accommodate a reduced level of growth but developing countries may not be able to sustain this reduction. This brings the discussion to international coordination and cooperation. We shall show in the next section the importance of international commitment for any policy measure initiated to reduce global warming to be successful.

2.3 Policy Coordination and International Cooperation

The recent surge in environmental consciousness in the industrialized countries has come about at a time when the majority of developing countries are concentrating their efforts on increasing the rate of economic growth. While developed countries are faced with environmental problems caused by development, the developing countries are faced with environmental problems caused by the very lack of development.

Therefore, an international protocol calling for a unilateral cut in carbon dioxide emissions will be met by stiff opposition from the developing countries. It will be difficult to persuade these countries to use more expensive carbon dioxide emission free energy sources to attain economic growth, especially when the developed countries themselves used cheap carbon dioxide emission intensive fuels to attain their present level of wealth. The difference in the hierarchical rankings of the economic and environmental objectives which exists between the developed countries on one hand and the developing countries on the other hand makes the process of drawing up an international protocol on carbon dioxide emissions complex.

To understand the complexity and the reasons for the existence of these complexities, it would be beneficial at this point to discuss briefly the role, developed and developing countries have played in the past and the role they will play in the future in the global warming problem. Rather than discussing each country's role in the global warming problem, the countries are divided into two groups.

Countries within each group are assumed to have identical policy objectives and are approximately at the same stage of industrialization. The first group is called the developed group and consists primarily the OECD countries as well as the Eastern European countries. The second group is called the developing group and includes China, India, South America, Africa, Southeast Asia, and the Middle East. Figure 2.2 below illustrates the present CO2 emissions by the various countries in both groups[2].

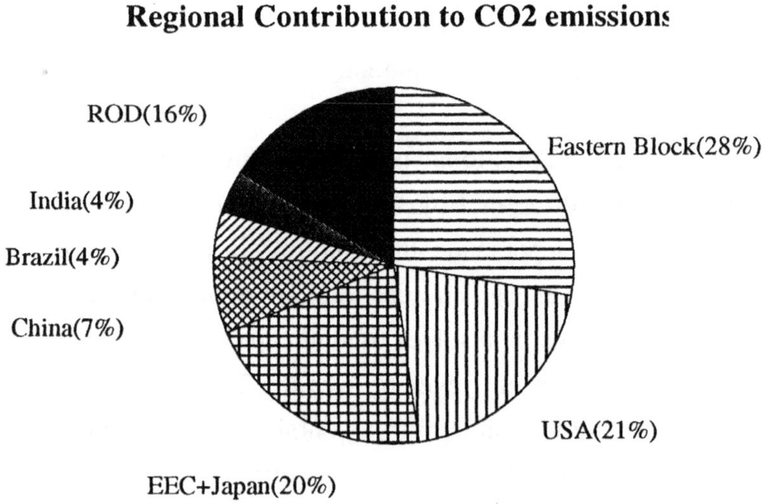

Figure 2.2 Regional CO2 emissions levels

[2]ROD stands for:- Rest of the Developing Countries

2.3.1 Developed countries and Global Warming: Past, Present, and Future.

Kellog (1981) estimates that the developed countries accounted for approximately 90 percent of global CO2 emissions in 1950. By 1976, the developed countries accounted for 80 percent of global emissions. The latest figures as shown in Figure 2.2 indicate that the developed countries share of global CO2 emissions is approximately 70 percent.

Although, since the mid 1970's, the developed countries have been able to reduce their use of fossil fuels and hence carbon dioxide emissions, Fuji (1990) asserts that the developed countries are primarily responsible for the realized 25 percent increase in atmospheric CO2 concentration. Therefore, we can safely assume for this study that the observed 80 parts per million volume (ppmv) increase in atmospheric carbon dioxide concentration since 1860 is attributed almost entirely to the developed countries.

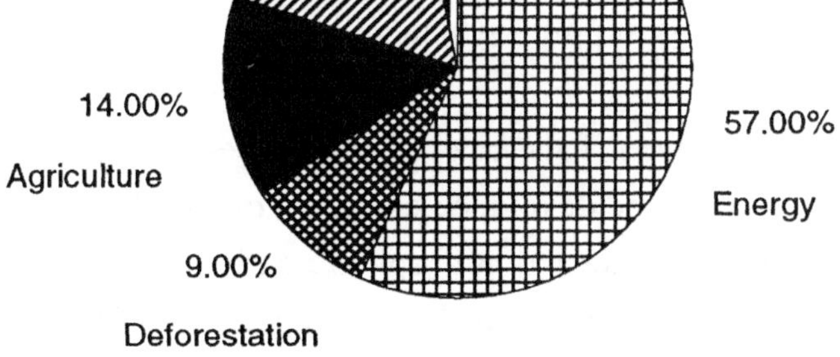

Fig 2.3 Contribution by sectors to the anthropogenic greenhouse effect.
Source: Lashof and Tirpak

Turning from a regional to a sectoral contribution perspective, Figure 2.3 shows that about 60 percent contribution towards global warming comes from the energy sector; and the energy sector contributes approximately 90 percent of CO_2 emissions. Before the 1974 oil shock, there was a high degree of correlation between economic growth and energy demand. But since the 1974 oil shock and the subsequent shocks, the rate of energy use has dropped even when real GNP was increasing in these countries.

The primary reason put forward to explain the reverse in energy use trend observed since the mid 1970's was the introduction of energy efficient processes which required less energy to produce the same level of output. The increase in the price of oil after the first oil shock caused the developed economies to shift to higher energy efficient processes. As most of these economies were operating at low energy efficiency levels, the marginal costs incurred by the higher energy efficient processes were attained at little or no costs to the economy (Jorgenson and Wilcoxen 1990). In fact, the trend of higher productivity due to use of more efficient processes was observed. However, it has been stated that marginal costs of energy efficiency improvements will increase exponentially as the level of efficiency increases. In other words, the first 20 percent in improving energy efficiency can be attained at little costs but the next 20 percent is more expensive and so on (Manne 1990). Furthermore, there is an upper limit to reducing CO_2 emissions by mere improvements in energy efficiency and once beyond this point, substitute energy supplies which do not emit CO_2 must be found.

Nevertheless, even if the rate of energy demand has been decreasing, the actual amount used by the developed countries due to the size of their economies, is sufficiently large to alter the atmospheric carbon dioxide concentration in the future. The present contribution of the developed countries towards global carbon dioxide anthropogenic emissions is about 70 percent. Therefore, even with the downward trend in energy use, the developed countries are still the major emitters of carbon dioxide. Furthermore, the size of these economies with respect to the size of population reflect an unusually large use of energy per capita. An average person in the U.S uses approximately 7.2 tons of petroleum equivalent per year versus a person who uses about 0.5 tons of petroleum equivalent per year in the developing countries.

This disproportionate size of economies with respect to the size of population is also a subject of debate when the issue of allocating emission rights is discussed. A small proportion of the world population has disproportionately used a large chunk of the globe's natural reservoir to absorb the extra emission of carbon dioxide. Therefore, in any future international agreement to limit carbon dioxide emissions, the developed countries will not only have to cut back further on the present level of energy use but also have to find some way of repaying the portion "borrowed" from the developing group's share. This is achieved in this study by incorporating financial aid flows to the developing group and at the same time cutting their own emissions to a level which is sustainable by the natural ecosystem.

2.3.2 Developing Group and Global Warming: Past, Present, and Future

The countries in this group have not played a predominant role in the global warming problem in the past but have been identified as major emitters in the future. As they embark on the same industrialization process the developed countries followed in the past, they are going to duplicate the carbon dioxide emission trends of the developed countries. But unlike the developed countries, the size of emissions is estimated to be ten fold larger due to the greater size population in these countries. This result arises from the assumption that the developing countries believe that energy is required for economic growth and base their objectives for economic prosperity on achieving the same energy per capita use as witnessed in the developed countries.

The present contribution by the developing group towards the global warming problem is about 29 percent. Of the 29 percent, 9 percent is from the burning of fossil fuels and the remaining 20 percent is from deforestation. Deforestation plays an important part when assessing the role developing countries play in global warming.

Deforestation occurs predominantly in the tropical rain forests belt. These forests provide a cheap source of cooking fuel and agricultural land for a large portion of the population in these countries who live below the poverty line. Unfortunately, tropical forests do not provide the best soils for agriculture and their usefulness as agricultural lands is short lived. These forests better serve man as a rich source of natural ecosystems which are crucial to the existence of life on this planet. Furthermore, tropical forests are important sinks for atmospheric carbon dioxide and play a major role in moderating the accumulation of carbon dioxide in the atmosphere. On the other hand, deforestation enhances the global warming problem because when trees are cut and burnt, large amounts of carbon dioxide are emitted.

The present rate of deforestation is approximately 11 million hectares per year (World Resources 1988) and if this trend does not decrease in the near future, most of the tropical forests on the globe are expected to disappear by the end of the next century. The planet as a single entity needs the tropical forests for its survival. Nevertheless, impoverished countries have no option other than letting its people cut these forests down to provide immediate relief to their fuel and food demands. Therefore, if the tropical forests are to be preserved, these countries must be given the financial resources to preserve these forests. It should be stressed that the preservation of these forests benefit all countries.

Why would these countries continue to emit carbon dioxide and in fact increase their emissions in the future? Once all the forest are cut down then their emission contribution will surely decline! However, as we explained earlier, the main objective of these countries is to industrialize and try to achieve the present living standards experienced in the developed countries. The only way they believe they can achieve this growth is by following the same path taken by the developed countries. Therefore, they would be reluctant to adopt any proposal which would hinder their primary objective. Furthermore, they could argue on the following grounds: why should they be subjected to a protocol when the developed countries were not restricted during their growth period?

From the discussion above, we can conclude that the developing countries will proceed on a path of industrialization to raise their living standards to the level presently experienced by the developed countries. The method they are expected to follow will be similar to the strategies taken by the developed countries. However, these strategies have been identified as the major source of greenhouse gas emissions. Therefore, if the developing countries follow the same path, the emission of greenhouse gases will increase exponentially. The relative size of the population in these regions will magnify the amount of greenhouse gases released by the developed countries. To put the magnitude of future CO2 emissions by the developing countries into perspective, it is estimated that China alone has the potential of emitting 750 gigatons (Gt) of carbon into the atmosphere if it uses its coal resources extensively in its drive to industrialization (Legget 1990). In comparison, the developed countries have, since 1860, put 175 Gt of carbon into the atmosphere.

The question that leaps to mind is: do the developing countries need to follow the same path the developed countries took to attain their present level of economic growth? This study addresses this question by exploring the possibility of leapfrogging the developing countries into the twenty first century by transferring new technology which is not energy intensive but is capable of helping them improve present standards of living. On the other hand, the developing countries will be willing to accept these new technologies only if they do not incur any additional costs in the growth process. However, new carbon dioxide emission free technologies in most cases cost more than the old carbon dioxide emission intensive technology. For example, China which has a large source of coal reserves will definitely opt to use this cheap source of energy rather than use more expensive oil, natural gas, or solar energy.

Therefore, any transfer of clean technology will need to be treated as emission reduction financial aid and not as a loan. This will give the developing countries an incentive to use these technologies. The issue which arises next is: why should the developed countries give financial aid to the developing countries especially if it is at a cost to their own economic growth? Especially if the developed countries are better equipped to buffer their economies from the immediate impacts of global warming than the developing countries. It would seem that the optimal solution for the developed countries would be to continue their present growth trends and prepare themselves for the impacts rather than use the funds as financial aid to the developing countries. This would be the optimal solution in the short run and if each individual economy considers itself as isolated from the actions of other economies.

However, we shall demonstrate later with our experiments that the adaptive solution on the part of the developed countries will cease to be optimal if: 1) the planning period is extended over a longer time horizon and 2) individual countries perceive their economic well being as linked to the economic well being of all other economies and that all economies operate within a closed system. Furthermore, from an ethical and moral point of view it would be unfair to future generations if present generations compromised the security of the planet and its continuing existence by their present actions.

2.4 Conclusion

The primary aim of this chapter was to highlight possible strategies, adaptive and preventative which could be used to tackle the global warming issue. However, the important issue in this chapter was the implementation of these policies in the different regions. The difference in the hierarchical rankings of economic and environmental objectives between the developed region on one hand and the developing regions on the other complicates the process of formulating an international protocol calling for the elimination of carbon dioxide. Some of the reasons for these differences were discussed and the conclusion which can be drawn is that for an international protocol to succeed, the cooperation of the developing regions is absolutely necessary. However, the task of ensuring the cooperation of these regions falls on the developed region. The primary method identified which can ensure cooperation is the transfer of twenty first century technology without subjecting the developing region to any additional costs. This is primarily achieved by allowing financial aid flows from the developed countries to the developing countries.

Chapter Three

The Mathematical Model

3.1 Introduction

The mathematical model formulated for this study can be classified under the broad category of optimal control models. An optimal control model has two unique characteristics which separate it from other model formulation styles. First, dynamic relationships are captured explicitly in the form of equations of motion (difference equations if in discrete time and differential equations if in continuous time). Second, the problem is set in an optimization framework.

Optimal control models can be solved either analytically or numerically. Analytical solution procedures are appropriate if the model is not too large or differentiated and the main objective is to analyze fundamental functional relationships within the model structure. However, the size restriction is too stringent for the model in this study and thus renders an analytical solution method inappropriate. This brings us to numerical methods. There are a number of solution algorithms within the context of numerical methods, each having its own advantages and disadvantages. The dynamic programming algorithm (Bellman 1968) comparative advantage lies in: (1) its ability to solve stochastic control models; and (2) its high level of efficiency in solving dynamic problems. However, the model structure imposed by the algorithm is relatively rigid; it is a non-trivial task to specify inequality constraints and multi-dimensional structures within the model framework. Non-linear programming techniques on the other hand allow the modeler a higher degree of freedom in the model formulation process. The structure is flexible to allow equality, inequality, identity, and multi-dimensional formats in the model framework. However, non-linear programming techniques do not take advantage of the dynamic structure as in the case of dynamic programming. Although the results are in no way effected, efficiency of the solution procedure is compromised. Furthermore, non-linear programming algorithms are not ideally suited to solve stochastic models.

The model used in this study is deterministic, multi-dimensional, and has a large number of inequality, equality, identity, and dynamic equations. Taking into consideration the pros and cons of the three methods discussed above, the non-linear programming solution method was determined as the "optimal" method to use in this study.

The optimization package used for formulating and solving the model using non-linear programming techniques is called the The General Algebraic Modeling System (GAMS) (Brooke, Kendrick, and Meeraus 1988). This package allows the modeler to translate the mathematical model into a format which is compatible with the non-linear algorithm in a simple and intuitive manner. The GAMS statement of the mathematical model is presented in Appendix One.

3.2 Model Nomenclature

In order to facilitate an easy and intuitive transformation process from the mathematical representation of the model to the model in GAMS syntax, a number of conversion protocols are adopted. The examples below will illustrate these protocols.

Examples

1). If a coefficient is represented mathematically as σ^{k}_{rsp}, then this variable is represented in GAMS by SIGMAK(R,S,P).

2). If a variable is represented mathematically as q^{p}_{rspt}, then this variable is represented in GAMS by QP(R,S,P,T).

In other words, the superscript becomes part of the name of the variable and the subscripts indicate the sets over which the variable is defined. The above notation style is also valid for equation and parameter specifications.

3.3 Set Specifications

The degree of aggregation in the model is represented by the set specification. In this study, there are four types of aggregation and therefore four set declarations. The aggregations are over space (regions), sectors, process, and time. Below is a summary of the set declarations.

R: Set of Regions

Deved:	Developed
Deveing:	Developing

S: Set of Sectors

Agri:	Agriculture
Industry:	Industry
Services:	Service

P: Set of Processes

Abat:	Abatement
Inter:	Intermediate
Inten:	Intensive

T: Set of Time Periods

1985, 1990, 1995, 2000, 2005, 2010, 2015, 2020, 2025, 2030, 2035 ;

The planning horizon spans a total of 50 years, with each time period in the model specification representing a 5 year interval.

3.4 The Model Equations

As mentioned in the introduction, the holistic model can be thought of as a system with two sub-systems; economic and climate. The economic system by itself includes 19 equation blocks (831 individual equations) and 16 variable blocks (1035 single variables). With the addition of the ecological sub-system (carbon cycle and temperature), the model expanded to 23 equation blocks (859 individual equations) and 19 variable blocks (1068 variables). The complete model, with the feedback effects had a total of 27 equation blocks (1136 individual equations) and 22 variable blocks (1343 single variables). The complete model, i.e., the holistic model therefore consists of 1136 individual equations and 1343 single variables.

Prices are not modeled explicitly in the model structure. The inclusion of prices would have increased the computational complexity considerably. A decision was made in the early stages of model development that although prices play an important role that its contribution in this particular study would be small when compared with the increase in computational complexity it adds to the model solving process. However, rather than completely ignore the role prices play, we decided to capture the effects of the price mechanism within the model structure in another manner without increasing the computational complexity.

The main reason prices are used in a majority of the present generation of global warming models is to capture price induced substitution effect between various fuel types to meet energy demand. In a majority of models, the prices of fuel types over the planning horizon are exogenously specified. Then emission target levels are specified. The models are than solved to obtain carbon tax rates which achieve these target levels. Due to the different carbon content of the different fossil fuels, depending on the emission target level and the forecasted prices, tax rates are determined. The final price, i.e., the forecast price plus the tax rate then determine the degree of substitution between the fuels to satisfy final energy demand in the economy. We decided to capture the substitution process through the use of emission protocols and the capital output coefficients. The former works in the following manner; for example, by placing a 50 percent emission protocol on CO2, the price of coal implicitly goes up with respect to the price of other fuels with a lower carbon content and it is this mechanism which causes substitution between the fuels. The capital-output coefficients on the other hand, capture the difference in the basic price structure between the fuel types. Although the price of the fuel is only part of the coefficient, we believe that by changing the capital-output coefficient over time, we capture the price change of major fuels to an approximation which is sufficient for the purposes of this study.

We do admit that this style of modeling the price mechanism is not as rigorous as a model structure with prices explicitly specified within the model structure as an endogenous variable. However, we should like to stress at this point that the main focus of the study is to identify the optimal level of CO2 emissions which gives the best trade-off position between economic performance and environmental damage. The inclusion of prices may have improved the magnitude resolution of the results but the qualitative results would have remained the same. However, we should like to point out that a majority of the models which use prices rely on forecasted fuel prices over the planning period. Forecasting fuel prices over a period of 50 years would just add another degree of uncertainty to a problem which is plagued by uncertainty.

The model structure is flexible for users to modify and add these price variables in the future. However, if the policy maker wants to capture the effect of dwindling oil and gas supplies, he or she can easily implement this by changing the capital-output coefficients appropriately over the planning horizon.

Figure 3.1 gives a schematic illustration of the model structure used in this study.

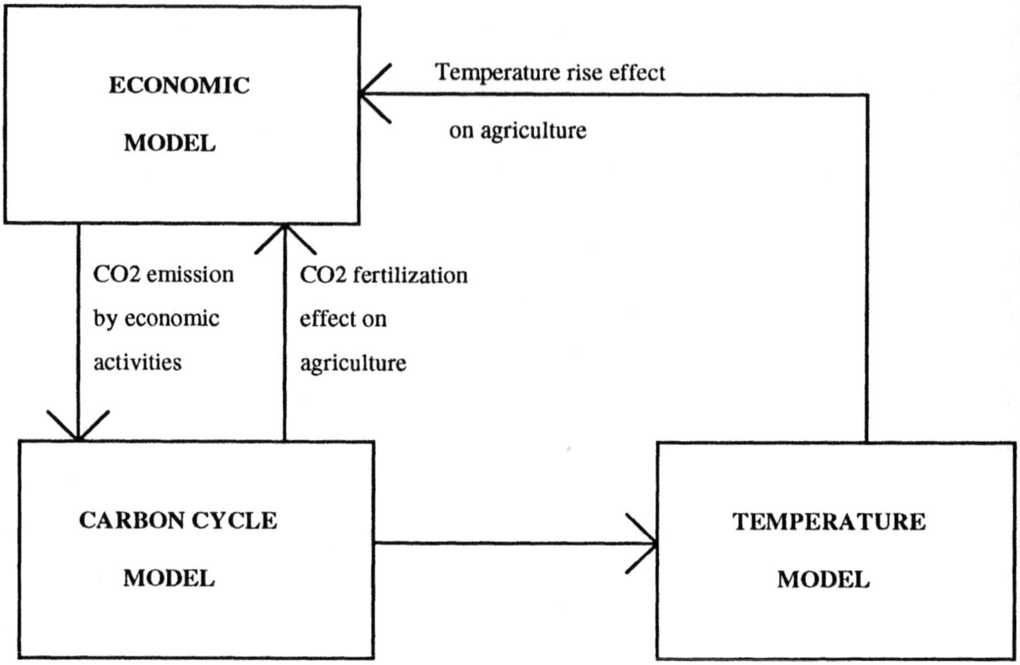

Figure 3.1 A schematic diagram of holistic model structure

The remaining part of this chapter will cover in detail the equations of the holistic model. The equations of the economic model will be discussed first followed by the feedback equations and finally the climate equations.

3.4.1 The Objective Function

Traditional macroeconomic planning models usually maximize a single utility function which represents social welfare. However, social welfare is a multi-dimensional phenomena (Hafkamp 1984) and is best illustrated by a set of indicators rather than a single indicator. Furthermore, these traditional planning models treat the planning process as a single agent optimization problem. This treatment is appropriate if the planning plan is for a single region but becomes inadequate for a multi-regional planning model in which there is more than a single decision maker. The planning model in this study has: (1) more than one optimizing decision maker; and (2) each decision maker has more than one desired objective and in most instances, these objectives are conflicting in nature. Therefore, the ideal objective function should be able to capture: (1) multi-objectives; (2) multi-agents; and (3) conflicting objectives.

The three criterion above are achieved by using a quadratic tracking function. A quadratic tracking function allows policymakers to state their economic and environmental objectives explicitly in the form of desired growth rates for important economic and environmental indicators. The policy objective is then to minimize the deviations of actual values from the desired levels over a period of time. Hence, the desired levels are the values that policymakers would like to achieve and the actual values the values that are eventually realized. Furthermore, the quadratic tracking function allows policymakers to assign a hierarchical structure of priorities on economic and ecological objectives by manipulating the weight structure in the priority matrix. The quadratic tracking function may be written as follows:

(1)
$$J = \frac{1}{2}\sum_{r\in R}\sum_{s\in S}\sum_{t\in T}\left[\left[w^{qr}_{rst}\left(z^{s}_{rst} - \tilde{z}^{s}_{rst}\right)^2\right] + \left[w^{c}_{rst}\left(c_{rst} - \tilde{c}_{rst}\right)^2\right]\right]$$
$$+\frac{1}{2}\sum_{r\in R}\sum_{t\in T}\left[\lambda^{f}_{rt}\left(f_{rt} - \tilde{f}_{rt}\right)^2\right] + \frac{1}{2}\sum_{r\in R}\sum_{s\in S}\sum_{t\in T}\left[\lambda^{i}_{rst}\left(i_{rst} - \tilde{i}_{rst}\right)^2\right]$$
$$+\frac{1}{2}\sum_{r\in R}\sum_{t\in T}\left[w^{e}_{rt}\left(e^{r}_{rt} - \tilde{e}^{r}_{rt}\right)^2\right]$$

w^{zs}_{rst} = priority weight on GDP in region r, sector s, and time t

z^{s}_{rst} = GDP in region r, sector s, and in time t.

$\tilde{z}^{s}_{rst}$ = desired path for GDP in region r, sector s, and in time t.

w^c_{rst} = priority weight on consumption variable in region r, sector s, and time t.
c_{rst} = consumption variable in region r, sector s, and in time t.
$\tilde{c}_{rst}$ = desired path for consumption level in region r, sector s, and in time t.

λ^f_{rt} = priority weight on deforestation variable in region r,and in time t.
f_{rt} = deforestation variable for region r and in time t.
$\tilde{f}_{rt}$ = desired path for deforestation level in region r and in time t.

λ^i_{rst} = priority weight on investment variable in region r, sector s, and in time t.
i_{rst} = investment variable in region r, sector s, and in time t.
$\tilde{i}_{rst}$ = desired path for investment levels in region r, sector s, and in time t.

w^e_{rt} = priority weight on atmospheric CO2 concentration by region r in time t.
e^r_{rt} = CO2 emission by region r in time t.
$\tilde{e}^r_{rt}$ = desired level of CO2 emission by region r in time t.

The quadratic tracking function above thus allows the policymakers to prioritize their policy objectives in an unique manner. The priority matrix in the objective function captures explicitly the degree of importance each policymaker places on a particular objective. Therefore, in our problem, if policymakers in a region place a higher priority on economic performance than on environmental integrity, this can easily be easily captured in the priority matrix by placing high weights on the economic variables with respect to the environmental variables.

3.4.2 Material Balance Constraint for Creditor Countries

The distribution relation is of the standard input-output style. This equation states that total sectorial output must be equal to the the sum of consumption demand, investment, intermediate requirements, and total foreign financial aid in a given point of time; i.e., sectorial output and demand must be in equilibrium. RD is a subset consisting of the donor countries, in this case, the developed region.

$$(2)\quad q^{s}_{rst} = \sum_{s'\in S} \alpha^{r}_{rss'} q^{s}_{rs't} + \sum_{s'\in S} \beta^{r}_{rss'} i_{rs't} + c_{rst} + h^{d}_{rst} \qquad r \in RD,\ s \in S,\ t \in T$$

$$[\text{output}] = \begin{bmatrix}\text{intermediate}\\ \text{requirements}\end{bmatrix} + [\text{investment}] + [\text{consumption}] + \begin{bmatrix}\text{aid to}\\ \text{developing region}\end{bmatrix}$$

$\alpha^{r}_{rss'}$ = input - output coefficient for region r.
$\beta^{r}_{rss'}$ = capital coefficient for region r.
h^{d}_{rst} = proportion of foreign aid contributed by sector s in region r in set of donar countries RD.

The proportion allocated to investment is determined by the beta matrix. The capital coefficient indicates the amount of resources which each sector contributes towards capital formation. The underlying assumption which makes it necessary to formulate investment demand as illustrated in equation 2 is that capital is not shiftable among the sectors.

3.4.3 Material Balance Constraint for the Debtor Countries

The balance equation is identical to Eq 2 except in the case of the recipient countries, foreign aid flow supplements the output capacity of the sector.

$$(3)\quad q^{s}_{rst} + h^{r}_{rst} = \sum_{s'\in S} \alpha^{r}_{rss'} q^{s}_{rs't} + \sum_{s'\in S} \beta^{r}_{rss'} i_{rs't} + c_{rst} \qquad r \in RR,\ s \in S,\ t \in T$$

$$[\text{output}] + \begin{bmatrix}\text{aid from}\\ \text{developed region}\end{bmatrix} = \begin{bmatrix}\text{intermediate}\\ \text{requirement}\end{bmatrix} + [\text{investment}] + [\text{consumption}]$$

h^r_{rst} = amount of foreign aid received by sector s by recipent region r in set RR which is a sub set of R.

3.4.4 Global Foreign Aid Balance Equation

This equation states that total foreign aid provided by the donor (developed) countries must be equal to the total foreign aid received by the recipient (developing) countries. In the material balance equations above, foreign aid sent and received is specified at the sectorial and regional level. If the global foreign aid balance equation is also specified at the regional and sectorial level, then aid sent by the agricultural sector in the developed region can be used only by the agricultural sector in the developing region and the same restriction holds for the other two sectors. Therefore, constraining aid at the sectorial level will give rise to inefficiencies in the aid collection and distribution process. Furthermore, formulating aid flows at the sectorial level may prove to be too restrictive and thus translate to the problem is having a smaller feasible set for an optimal solution.

However, by summing over the sectors, this stringent restriction is prevented. Toal aid from all sectors in the developed region is collected into a pool from which all three sectors in the developing region can have access to.

$$(4)\quad \sum_{s\in S}\sum_{r\in RR} h^r_{rst} = \sum_{s\in S}\sum_{r\in RD} h^d_{rst} \qquad t \in T$$

$$\begin{bmatrix}\text{total aid received by} \\ \text{developing region}\end{bmatrix} = \begin{bmatrix}\text{total aid sent by} \\ \text{developed region}\end{bmatrix}$$

3.4.5 Foreign Aid Condition Constraint

The mechanism by which the developing countries are "leapfrogged" into an era of modern technology is financial aid flows. However, to ensure that aid received from the developed countries is directed towards the adoption of new and clean technology, the level of aid is tied to the level of output produced by the new technologies in the previous time period. This formulation provides motivation on the part of the developing countries to use pollution abatement and intermediate processes; i.e., the anticipation of future aid influences present investment decisions on the part of the developing countries. In other words the level of aid from the developed region is dependent on ex-ante investment decisions by the developing region.

However, rather than requiring the developed countries to finance 100 percent of the new technologies, the formulation in this study assumes that the developed region will finance only the cost differential experienced by the developing region due to the adoption of the more expensive but less CO2 emitting technology. For example, China chooses the option to use a combination of purified coal, gas and solar technology rather than just use high CO2 emission coal processes. Then, under the scenario used in this study, the developed region will finance the difference between what it would have cost China to use the old coal technology versus the costs incurred from the environment sensitive mix of technology.

To maintain simplicity and to keep the initial model small, aid is restricted to cost differentials incurred by the use of extra capital. Land is not part of the aid deal as the recipient countries are assumed to have a comparative advantage in the land intensive pollution abatement process. But the equation formulation allows this assumption to be relaxed easily in the future to allow the possibility of aid tied to the preservation of forest lands.

(5) $h^{r}_{rst} = \left(\mu^{k}_{rsp^1} - \mu^{k}_{rsp^3}\right) q^{p}_{rsp^1,t-1}$

$$\begin{bmatrix}\text{aid received by}\\ \text{developing region}\end{bmatrix} = \begin{bmatrix}\text{capital output cost difference}\\ \text{between abatement and}\\ \text{intensive processes}\end{bmatrix}\begin{bmatrix}\text{output produced using}\\ \text{abatement process in}\\ \text{previous time period}\end{bmatrix}$$

$+\left(\mu^{k}_{rsp^2} - \mu^{k}_{rsp^3}\right) q^{p}_{rsp^2,t-1}$ $\qquad r \in RR$

$s \in S$

$t \in T$

$$+\begin{bmatrix}\text{capital output cost difference}\\ \text{between intermediate and}\\ \text{intensive processes}\end{bmatrix}\begin{bmatrix}\text{output produced using}\\ \text{intermediate process in}\\ \text{previous time period}\end{bmatrix}$$

3.4.6 Capital Accumulation

A linear first-order difference equation is formulated to capture the capital accumulation process. There are two methods of modelling capital use and accumulation, capital by sector of origin and capital by sector of use. Capital by sector of origin denotes capital as shiftable among sectors whereas capital by sector of use assumes that capital is not shiftable among sectors but is unique for each sector. The formulations used in equations 6, 2a, and 3 assume that capital is not shiftable among sectors but is shiftable among the three processes within each sector. In other words, capital is treated as capital by sector of use.

(6) $k_{rs,t+1} = (1-\delta_{rs})k_{rst} + \tau\, i_{rst}$ $\qquad r \in R$

$s \in S$

$t \in T$

$$\begin{bmatrix}\text{capital stock at start}\\ \text{of time period t+1}\end{bmatrix} = \begin{bmatrix}\text{1 - capital}\\ \text{consumption rate}\end{bmatrix}\begin{bmatrix}\text{capital stock at start}\\ \text{of time period t}\end{bmatrix} + \begin{bmatrix}\text{investment in}\\ \text{time period t}\end{bmatrix}$$

k_{rst} = capital stock in region r, sector s, and in time t.

δ_{rs} = depreciation rate in region r and for sector s.

τ = number of years per time period.

3.4.7 Capital Constraint

The capital stock constraint indirectly defines the production function. A Leontief style production function is used but with substitution possibilities from alternate processes. Also there are modifications to capture the feedback effects of carbon dioxide concentration and temperature change. We begin by defining three different processes, each using different proportions of inputs. Two inputs are identified for production purposes; capital and land.[1] Equation 7 states that the sum of capital used by all three processes cannot exceed the total amount of capital available. As mentioned before, capital is shiftable within the processes but not across sectors (capital by sector of use).

$$(7) \quad \sum_{p \in P} \kappa_{rspt} q^{p}_{rspt} \leq k_{rst} \qquad r \in R, \; s \in S, \; t \in T$$

$$\begin{bmatrix}\text{capital output} \\ \text{coefficient}\end{bmatrix} \begin{bmatrix}\text{sectoral} \\ \text{output}\end{bmatrix} \leq \begin{bmatrix}\text{capital stock} \\ \text{in sector}\end{bmatrix}$$

κ_{rspt} = The number of units of capital required in region r and sector s by process p to produce one unit of output.

q^{p}_{rspt} = production level in region r, sector s, by process p and in time t.

---

[1]Labor is not considered in this study. It will be an appropriate extension for future work.

3.4.8 Land Constraint

Equation 8 states that the total amount of land which can be used for production[2] cannot exceed the available supply. It is assumed that land is used primarily for agricultural activities. Therefore, the land constraint is solely determined by the level of agricultural activity.

$$(8) \quad \sum_{s \in S} \sum_{p \in P} \pi^{d}_{rsp} q^{p}_{rspt} \leq d_{rt} \qquad r \in R \quad t \in T$$

$$\begin{bmatrix}\text{land output} \\ \text{coefficient}\end{bmatrix}\begin{bmatrix}\text{output level at} \\ \text{process level}\end{bmatrix} \leq \begin{bmatrix}\text{supply of} \\ \text{land}\end{bmatrix}$$

π^{d}_{rsp} = number of land units required in region r and sector s by process p to produce one unit of output.

d_{rt} = level of land units available in region r at time t.

3.4.9 Regional Sectorial Production Level

To maintain consistency with the material balance equation, a variable denoting sectorial output in each region is introduced. Therefore, regional sectorial production level is the sum of output from the three processes.

$$(9) \quad q^{s}_{rst} = \sum_{p \in P} q^{p}_{rspt} \qquad r \in R \quad s \in S \quad t \in T$$

$$\begin{bmatrix}\text{sectoral output} \\ \text{level}\end{bmatrix} = \begin{bmatrix}\text{sum of outputs by} \\ \text{all processes in sector}\end{bmatrix}$$

---

[2] We assume that only the agriculture sector uses land as an input in the production function and that the energy sector does not.

3.4.10 Capital-Output Coefficient

In traditional planning models, a Leontief type fixed capital-output production functional form is used. In more sophisticated non-linear models, Cobb-Douglas or CES type production functions are specified to illustrate the production process. However, in this study, the main emphasis is on capturing the feedback effects of environmental changes on the economic production system. But as there is incomplete empirical evidence on the feedback effects, the statistical estimation of a Cobb-Douglas or a CES production function incorporating the feedback effects was deemed as inadequate.

Therefore, to keep the formulation as simple as possible but at the same time capture the feedback effects, the fixed capital-output Leontief type production function formulation is modified to allow the capital-output ratio to be a function of atmospheric CO2 concentration and temperature rise. The primary assumption underlying the formulation is that as environmental variables change, policymakers respond to the change by increasing inputs, i.e., capital and land to maintain desired output levels in the agriculture sector. The policy response group of the Intergovernmental Panel on Climate Change (IPCC 1990) concluded that climate change would not reduce global food output because of adaptation. However, the cost of maintaining output levels increases due because of the implementation of these adaptation strategies in response to climate changes. We capture these costs explicitly via the feedback equations.

(10) $\kappa_{rsp,t+1} \quad = \quad \mu^{k}_{rsp} \quad + \quad v^{k}_{rsp}\left(t_{t+1} - t_{1985}\right)$

$$\begin{bmatrix}\text{capital output}\\ \text{coefficient}\end{bmatrix} = \begin{bmatrix}\text{fixed capital output}\\ \text{coefficient}\end{bmatrix} + \begin{bmatrix}\text{temperature rise}\\ \text{coefficient}\end{bmatrix}\begin{bmatrix}\text{change in}\\ \text{temperature}\end{bmatrix}$$

$+ \quad \sigma^{k}_{rsp}\left(n^{a}_{t+1} - n^{a}_{1985}\right)$ $\qquad r \in R$

$s \in S$

$p \in P$

$t \in T$

$$+\begin{bmatrix}\text{CO2 change}\\ \text{coefficient}\end{bmatrix}\begin{bmatrix}\text{change in CO2}\\ \text{concentrations}\end{bmatrix}$$

v^{k}_{rsp} = coefficient reflecting the effect of unit rise intemperature on capital units required to produce one unit of output in sectors.

σ^{k}_{rsp} = coefficient reflecting the effect of unit rise in atmospheric CO2 concentration on number of capital units required to produce unit output in sectors.

As equation 10 illustrates, the capital-output coefficient is composed of a constant term which is the normal capital-output coefficient used in lieu of environmental feedback effects. The v^{k}_{rsp} and σ^{k}_{rsp} coefficients reflect the effect of a unit rise in temperature and CO2 concentration on the capital-output coefficients respectively. For example, if no temperature or carbon dioxide concentration rises are experienced, then the number of units of capital required to produce one unit of output is μ^{k}_{rsp}.

However, if temperature rises by a degree centigrade, then agricultural output is expected to drop. Policymakers in the regions are expected to react to this drop by increasing the capital input so as to prevent the drop in output; i.e., the capital-output coefficient increases. The v^{k}_{rsp} value reflects the impact of temperature increase on irrigation expenditures, soil enhancement programs, and the use of additional specialized machinery use in the agricultural sector to adapt to the new environment[3]. The σ^{k}_{rsp} parameter reflects the impact of increased levels of carbon dioxide concentration on the yields of crops[4]. The parameters v^{k}_{rsp} and σ^{k}_{rsp} are still at the stage of debate and the calculations used to derive these parameters are based on estimations prepared by the United Nations Environmental Programme (UNEP 1987) and the Central Intelligence Agency (CIA 1978).

3.4.11 Land-Output Coefficient

The land-output coefficient expressed in equation 11 is similar to the capital-output equation expressed in Equation 10.

$$\textbf{(11)} \quad \pi_{rsp,t+1} = \mu^{d}_{rsp} + v^{d}_{rsp}\left(t_{t+1} - t_{1985}\right)$$

$$\begin{bmatrix}\text{land output}\\ \text{coefficient}\\ \text{in period } t+1\end{bmatrix} = \begin{bmatrix}\text{fixed land}\\ \text{output}\\ \text{coefficient}\end{bmatrix} + \begin{bmatrix}\text{temperature}\\ \text{rise}\\ \text{coefficient}\end{bmatrix}\begin{bmatrix}\text{change in}\\ \text{temperature from}\\ \text{base year}\end{bmatrix}$$

$$+ \quad \sigma^{d}_{rsp}\left(n^{a}_{t+1} - n^{a}_{1985}\right) \qquad r \in R,\; s \in S,\; p \in P,\; t \in T$$

$$+\begin{bmatrix}\text{CO2 change}\\ \text{coefficient}\end{bmatrix}\begin{bmatrix}\text{change in atmospheric}\\ \text{CO2 concentrations}\\ \text{from base year}\end{bmatrix}$$

---

[3]Thompson,L.M. "Weather variability, climate change, and grain production". Science 189, 1978, pp 535-541

[4]United Nations Environment Programme. The Changing Atmosphere. Nairobi, Kenya. UNEP Environmental Brief No1, 1988.

v^{d}_{rsp} = coefficient reflecting the effect of unit temperature rise on the number of capital units required by process p toproduce one unit of output in secto and in region r.

σ^{d}_{rsp} = coefficient reflecting the effect of unit rise inatmospheric CO2 concentration on the number of land units required by process produce unit output in sector s and in region r.

As in the case of capital, the parameter μ^{d}_{rsp} is identical to the fixed land coefficient used in the Equation 5. The v^{d}_{rsp} coefficient reflects the effect a global rise in temperature will have on land productivity; higher temperatures result in higher evaporation rates, thus causing a drop in land productivity.[5]. This translates to larger amounts of land required to maintain present levels of production. The σ^{d}_{rsp} coefficient reflects the effect of higher yields stemming from higher concentration levels of carbon dioxide which will increase the productivity of land. This effect will diminish the effect of σ^{d}_{rsp} to a certain extent. The extent of this difference between the parameters varies between regions and is crucial in marginal agricultural regions; primarily in the developing regions.

3.4.12 Supply of Land

A linear first order difference equation is used to model land accumulation. Equation 11 states that the level of land available in time t+1 is equal to the amount of land available in time t plus the amount of land made available through deforestation[6] in time t. There are a number of assumptions associated with this formulation. First, the study ignores reforestation and second, all deforested land is assumed to be used for agricultural activities.

---

[5]Parry,H.M. Climate Change, Agriculture & Settlement. Folkstone, England: Dawson, 1978.
[6]We assume that all land made available through deforestation can be used for agriculture.

$$(12)\quad d_{r,t+1} \quad = \quad d_{rt} \quad + \quad \tau f_{rt} \qquad r \in R,\ t \in T$$

$$\begin{bmatrix}\text{supply of land at}\\ \text{beginning of time}\\ \text{period } t+1\end{bmatrix} = \begin{bmatrix}\text{supply of land at}\\ \text{beginning of time}\\ \text{period } t\end{bmatrix} + \begin{bmatrix}\text{forest deforested}\\ \text{in time period } t\end{bmatrix}$$

f_{rt} = deforestation rate in region r and in time t.

3.4.13 Deforestation Constraint

This constraint is required in the solution process if an unbounded solution is to be avoided. Deforestation is a control variable and is left at the discretion of the policymaker. Therefore, if there isn't an upper bound, the solution process will be indeterminate and the problem becomes unbounded.

$$(13)\quad \sum_{t \in T} f_{rt} \quad \leq \quad b_r^o \qquad r \in R$$

$$\begin{bmatrix}\text{deforestation}\\ \text{level}\end{bmatrix} \leq \begin{bmatrix}\text{total supply}\\ \text{of forest land}\end{bmatrix}$$

b_r^o = amount of forest land in base year in region r.

We have adopted a relaxed upperbound by allowing policymakers the option of cutting all forest land within the planning period.

3.4.14 Energy Demand

Once the level of output by each process in the service and industry sectors has been determined, the next step is to identify the energy demanded by this level of output. This is done by multiplying the level of output by the energy per unit output coefficient. This coefficient shows the amount of energy in exajoules which is required for a dollar output in the respective sectors.

(14) $q^{j}_{rspt} = \zeta^{j}_{rs} q^{p}_{rspt}$ $r \in R$, $s \in S$, $p \in P$, $t \in T$

$$\begin{bmatrix}\text{energy demand}\\ \text{in exajoules}\end{bmatrix} = \begin{bmatrix}\text{energy output}\\ \text{coefficient}\end{bmatrix} \begin{bmatrix}\text{output at}\\ \text{process level}\end{bmatrix}$$

ζ^{j}_{s} = exajoules per dollar output in sector s

3.4.15 Emission of CO2 by Fossil Burning

The level of CO2 emission by fossil fuels is dependent on the level of output using the three different processes in the sectors. The coefficients ζ_p below are technical parameters calculated from the physical properties of the coal, oil, and natural gas.

(15) $e_{rspt} = \zeta_{p} q^{j}_{rspt}$ $r \in R$, $s \in S$, $p \in P$, $t \in T$

$$\begin{bmatrix}\text{CO2 emission by}\\ \text{fossil fuel burning}\end{bmatrix} = \begin{bmatrix}\text{CO2 emission}\\ \text{coefficient}\end{bmatrix} \begin{bmatrix}\text{amount of exajoules}\\ \text{provided by respective}\\ \text{process}\end{bmatrix}$$

e_{rspt} = emission level in region r, sector s, by process p, and in time t.
ζ_p = amount of CO2 emitted by unit level of energy unit by process p.

3.4.16 Regional pollution level

The regional pollution level illustrates the level of carbon dioxide each region emits in each time period. There are two primary sources of CO2 emissions in this model- burning of fossil fuels and deforestation.

$$(16) \quad e^{r}_{rt} = \sum_{s\in S}\sum_{p\in P} e_{rspt} + \zeta^{f}_{r} f_{rt} \qquad r \in R,\ t \in T$$

$$\begin{bmatrix}\text{level of CO2}\\ \text{emission}\end{bmatrix} = \begin{bmatrix}\text{emission of CO2 by}\\ \text{fossil fuel burning}\end{bmatrix} + \begin{bmatrix}\text{emission of CO2 by}\\ \text{deforestation}\end{bmatrix}$$

e^{r}_{rt} = pollution emission level by region r and in time t.
ζ^{f}_{r} = pollution emission coefficient by deforestation in region r and in time t.

3.4.17 Global CO2 Emissions

The link between the economic sub-system and the ecological sub-system takes the form of the global CO2 emissions, Global CO2 emissions is the sum of each region's CO2 emissions.

$$(17) \quad e^{g}_{t} = \sum_{r} e^{r}_{rt} \qquad t \in T$$

$$\begin{bmatrix}\text{global CO2}\\ \text{emission}\end{bmatrix} = \begin{bmatrix}\text{sum of regional}\\ \text{CO2 emissions}\end{bmatrix}$$

e^{g}_{t} = global pollution emission level in time t.

3.4.18 Sectorial Gross Domestic Product

$$(18) \quad z_{rst}^{s} = q_{rst}^{s} - \sum_{s'} \alpha_{rss'}^{r} q_{rs't}^{s} \qquad r \in R, \; s \in S, \; t \in T$$

$$\begin{bmatrix}\text{GDP per} \\ \text{sector}\end{bmatrix} = \begin{bmatrix}\text{output per} \\ \text{sector}\end{bmatrix} - \begin{bmatrix}\text{intermediate} \\ \text{requirement}\end{bmatrix}$$

3.4.19 Regional Gross Domestic Product

$$(19) \quad z_{rt}^{r} = \sum_{s \in S} z_{rst}^{s} \qquad r \in R, \; t \in T$$

$$\begin{bmatrix}\text{regional} \\ \text{GDP}\end{bmatrix} = \begin{bmatrix}\text{sum of sectoral} \\ \text{GDP}\end{bmatrix}$$

3.4.20 Carbon Cycle Equations

The amount of carbon dioxide emitted by human activities is not the only source of carbon dioxide. The ocean, atmosphere, and land act as sources and sinks for carbon dioxide. These three "reservoirs" can be further sub-divided in order to capture the movements among the reservoirs in more detail. The degree of division is correlated with the extent of the study. However, we believe that the 3 reservoir model of Keeling (1979) shown in Figure 3.2 will suffice for our needs.

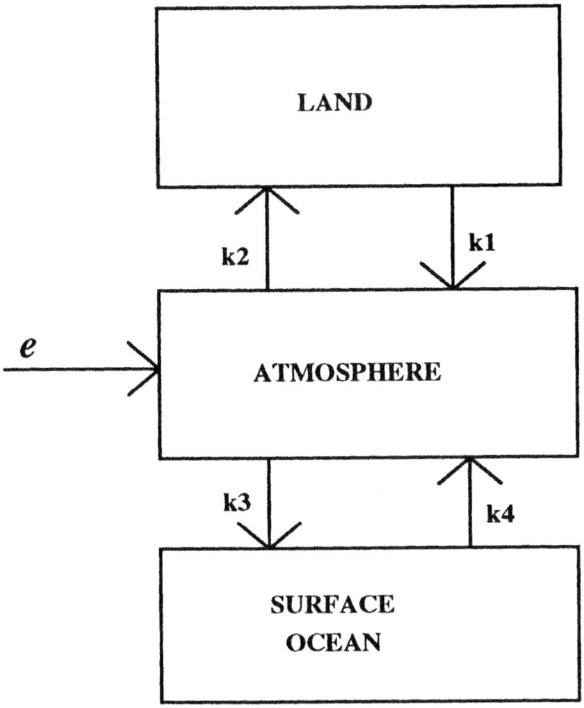

Figure 3.2. A carbon cycle model.

(20a) $n^{b}_{t+1} = k_2 n^{a}_{t}$

$$\begin{bmatrix}\text{CO2 concentration}\\ \text{in biosphere at}\\ \text{beginning of time t+1}\end{bmatrix} = \begin{bmatrix}\text{CO2 transfer}\\ \text{coefficient from}\\ \text{atmosphere to biosphere}\end{bmatrix}\begin{bmatrix}\text{CO2 concentration}\\ \text{in atmosphere at}\\ \text{beginning of time t}\end{bmatrix}$$

$+ \quad (1-k_1)n^{b}_{t} \qquad t \in T$

$$+\begin{bmatrix}\text{1-CO2 transfer}\\ \text{coefficient from}\\ \text{biosphere to atmosphere}\end{bmatrix}\begin{bmatrix}\text{CO2 concentration}\\ \text{in biosphere at}\\ \text{beginning of time t}\end{bmatrix}$$

$$\textbf{(20b)} \quad n^{a}_{t+1} = e^{g}_{t} + (1-k_2-k_3)n^{a}_{t}$$

$$\begin{bmatrix}\text{CO2 concentration}\\ \text{in atmosphere at}\\ \text{beginning of}\\ \text{time t+1}\end{bmatrix} = \begin{bmatrix}\text{global}\\ \text{CO2}\\ \text{emissions}\end{bmatrix} + \begin{bmatrix}\text{1 - sum of CO2 transfer}\\ \text{coefficients from}\\ \text{atmosphere to other}\\ \text{reservoirs}\end{bmatrix}\begin{bmatrix}\text{CO2}\\ \text{concentration}\\ \text{in atmosphere}\\ \text{at beginning}\\ \text{of time t}\end{bmatrix}$$

$$+ \quad k_1 n^{b}_{t} \quad + \quad k_4 n^{m}_{t} \qquad t \in T$$

$$+\begin{bmatrix}\text{CO2 transfer}\\ \text{coefficent}\\ \text{from biosphere to}\\ \text{atmosphere}\end{bmatrix}\begin{bmatrix}\text{CO2 concentration}\\ \text{in biosphere at}\\ \text{beginning of time t}\end{bmatrix} + \begin{bmatrix}\text{CO2 transfer}\\ \text{coefficient}\\ \text{from ocean to}\\ \text{atmosphere}\end{bmatrix}\begin{bmatrix}\text{CO2}\\ \text{concentration}\\ \text{in ocean at}\\ \text{begining}\\ \text{of time t}\end{bmatrix}$$

$$\textbf{(20c)} \quad n^{m}_{t+1} = k_3 n^{a}_{t}$$

$$\begin{bmatrix}\text{CO2 concentration in}\\ \text{ocean at beginning of}\\ \text{time t+1}\end{bmatrix} = \begin{bmatrix}\text{CO2 transfer coefficient}\\ \text{from atmosphere to}\\ \text{ocean}\end{bmatrix}\begin{bmatrix}\text{CO2 concentration in}\\ \text{atmosphere at}\\ \text{beginning of time t}\end{bmatrix}$$

$$+ \quad (1-k_4)n^{m}_{t} \qquad t \in T$$

$$+\begin{bmatrix}\text{1 - CO2 transfer coefficient}\\ \text{from ocean to atmosphere}\end{bmatrix}\begin{bmatrix}\text{CO2 concentration in}\\ \text{ocean at beginning}\\ \text{of time t}\end{bmatrix}$$

n_t^b = Carbon mass in biota in time t.
n_t^a = Carbon mass in the atmosphere at time t.
n_t^m = Carbon mass in surface ocean in time t.

k_1 = transfer coefficient for CO2 from land to atmosphere.
k_2 = transfer coefficient for CO2 from atmosphere to land.
k_3 = transfer coefficient for CO2 from atmosphere to surface ocean
k_4 = transfer coefficient for CO2 from surface ocean to atmosphere.

The k parameters are called transfer coefficients which illustrate the rate of exchange of CO2 between the reservoirs. The k coefficients are assumed to be time invariant. *n* is the level of carbon mass in Gt at the respective reservoirs (n_a=atmosphere, n_b=land, n_m=ocean). The e variable is the influx of carbon dioxide emission by economic activities; global CO2 emissions.

3.4.21 Initial Conditions for Atmospheric CO2 concentration

For the purpose of this study, we shall concentrate on the actual changes in the carbon mass in each reservoir. We therefore assume that at time zero, the system is in steady state as illustrated by equation 21 below.

(21)
$$n_t^a = 760 \text{ Gt}$$
$$n_t^b = 2060 \text{ Gt}$$
$$n_t^m = 1005 \text{ Gt} \qquad t \in \{1985\}$$

3.4.22 Conversion Formula from Mass to Concentration

Equations 20 and 21 determine the carbon mass in the atmosphere. The next step is to find the carbon concentration in the atmosphere. From physical properties of carbon dioxide as well as the atmosphere, we derive the conversion factor of 2.12[7].

$$(22) \quad e_t^{rpmv} = \frac{n_t^a}{2.12} \qquad t \in T$$

3.4.23 The Global Temperature Equation

Modeling global temperature changes caused by increases in atmospheric CO2 concentration is a complex and difficult task. The most sophisticated models which predict temperature change on a regional and seasonal level take weeks to solve on a supercomputer. A simplified model which predicts the average global surface temperature based on the atmospheric concentration level of CO2 is used in this study. Therefore, if there is a doubling of atmospheric CO2 concentration, then all countries are assumed to experience the same level of temperature change. This assumption may be too restrictive for a detailed analysis of the global warming problem, but it is considered adequate within the context of the study objective. A logarithmic function is used to denote the functional relationship between temperature change and atmospheric CO2 concentration because the absorption effect of CO2 decreases as the concentration levels increase (Masters 1991). Equation 22 is a logarithmic function which gives a globally averaged surface temperature change due to an increase in atmospheric CO2 concentrations.

---

[7]The reader is requested to refer to the IPCC Scientific Assessment report, pg 9, for a detailed computation.

$$(23)\ \Delta t_t = \frac{\Delta t_d}{\ln 2} \ln \left[\frac{e_t^{rpmv}}{e_{"base"}^{rpmv}} \right] \qquad t \in T$$

$$\begin{bmatrix}\text{change in surface} \\ \text{temperature at time t}\end{bmatrix} = \begin{bmatrix}\text{constant term} \\ \hline \text{constant term}\end{bmatrix} \ln \begin{bmatrix} \dfrac{\text{CO2 concentration in time t}}{\text{CO2 concentration in base period}} \end{bmatrix}$$

Δt_t = equilibrium global temperature change

Δt_d = equilibrium temperature change predicted if CO2 concentration doubles.

n_t^a = atmospheric CO2 concentration at time t.

$n_{"base"}^a$ = atmospheric CO2 concentration at steady state of 270 ppmv.

An atmospheric CO2 concentration of 270 ppmv is assumed to be the steady state concentration level in which there is no change in temperature. Therefore, any change in atmospheric CO2 concentration from the steady state is assumed to cause a change in surface temperatures. For example, the atmospheric CO2 concentration in 1985 is approximately 354 ppmv. According to equation 22, the rise in surface temperature will be approximately 0.3 degrees Celsius. But as the change is from a steady state of zero, the term Δt_t is used synonymously with t_t in the model.

This chapter has dealt primarily with the mathematical formulations governing the model structure. The GAMS statement of the complete model is presented in Appendix One. The data used for the model is presented in the next chapter and the coefficients derived for the feedback equations are presented in detail.

Chapter Four

Numerical Data for a Multi-Regional, Multi-Sectorial, and Multi-Process Optimal Growth Model

4.1 Introduction

The multi-country inter-sectoral style modeling of global warming is still at an infancy stage. This implies that there are no uniform and easily accessible databases from which all the relationships of the model can be econometrically estimated. Rather, it is necessary to bring together disparate data from many sources and fashion them into a model which captures the essential elements of the problem. Such a model can then be solved to provide insights to crucial relationships between economic activities and the enhanced greenhouse effect. The solution along with an accompanying sensitivity analysis study can then be used to identify the data which are crucial to the outcome of the model results. These data can then be further studied in detail and the relationships involving them can be estimated to improve the quality of the results.

However, if these relationships still cannot be estimated due to the lack of data, then plausible ranges of the crucial parameters which have significant effects on the results can be highlighted. This is especially important because the projected temperature increases caused by global warming have yet to be realized which means that the feedback effects have not materialized. Therefore, a regression analysis done between temperature changes and agricultural yields would not capture any significant relationship between the two at the present.

Due to the multi-disciplinary nature of the study, data was collected from a large number of sources. Therefore, a considerable effort was made to ensure the consistency of the data and the model parameters. The model parameters can be categorized into two groups. The first group contains parameters adopted from previous studies while the second group consists of parameters computed within this study. The remaining part of the chapter will cover in detail parameters from both groups.

4.2 Input-output coefficients (A Matrix)

The input-output coefficients[1] used in this study were adopted from the study done by Bagchi (1984). The sectorial classification used in the Bagchi model was in close agreement with the sectorial classification used in this study- agriculture, industry, and services. The agriculture sector was composed of sectors producing agricultural and food products. The industry sector was composed of sectors producing chemicals, equipment and metals. The services sector was composed of transport, communication, construction and services. Table 4.1 contains the coefficients for the developed region while Table 4.2 illustrates coefficients used in the developing region.

Sector of origin	Sector of Destination		
	Agriculture	Industry	Service
Agriculture	0.34	0.05	0.03
Industry	0.08	.33	.11
Service	.12	.14	.20

Table 4.1 Input-Output coefficients for the developed region

[1]These coefficients are illustrated in Table Alphar in the GAMS statement.

Sector of origin	Sector of Destination		
	Agriculture	Industry	Service
Agriculture	0.17	0.10	0.01
Industry	0.07	.30	.07
Service	.14	.18	.14

Table 4.2 Input-Output coefficients for the developing region

Representative values are used to illustrate the input-output coefficients for the regions. Therefore, the input-output coefficients in Table 4.1 were adopted from an input-output table for the EEC (Bagchi 1984 pg 169). An input-output table for India (ibid) is used to represent the input-output coefficients for the developing region.

4.3 Capital Coefficients (B Matrix)

The capital coefficients[2] in Table 4.3 show the proportion of output each sector contributes for investment purposes.

Sector of origin	Sector of Destination		
	Agriculture	Industry	Service
Agriculture	0.0	0.0	0.0
Industry	0.69	1.39	0.35
Service	0.0	0.0	0.0

Table 4.3. Capital coefficients for both regions.

Not all three sectors contribute to capital formation (Kendrick and Taylor 1978 pg 234). It is assumed that the industry sector is the primary sector which contributes to capital formation for itself as well for the other two sectors. These coefficients were adopted from a study conducted by Kendrick and Taylor (ibid) on the Korean economy. Although the coefficients are representative of the Korean economy (developing), we use the coefficients to represent both the developed and developing regions. This assumption is made on the basis that irrespective of the regions, the industrialization process is the same for both regions and it will be the same sectors which are responsible for capital formation. However, the proportion of resources which are diverted from the

[2]These are presented in Table Beta in the GAMS statement.

industrial sector to the other sectors for investment may differ across regions; therefore, the actual value of the coefficients may differ.

Since estimation of these coefficients is a non-trivial task and as we believe that a change in these capital coefficients would not change the general properties of the model, it was deemed appropriate to use one set of coefficients for both regions. The change in capital coefficients would have changed the base results of the model but results from subsequent experiments would also reflect the change and though there may be a difference in the optimal values of the various variables in the model, the pattern of the results from the experiment would be the same. As noted from previous chapters, this study is primarily interested in evaluating the efficiency criteria among a number of different policy experiments and not in forecasting economic indicators. Therefore, the assumption of the same capital coefficients for the two regions will not significantly affect the primary objective of this study.

4.4 Capital-Output Coefficient

The capital-output coefficient is a variable dependent on three factors: (1) a constant term which illustrates the level of capital required to produce a single unit of output in lieu of environmental feedbacks; (2) a second term which captures the effect of temperature changes on output level and thus the number of capital units; and (3) a third term which reflects the CO2 fertilization effect on agricultural output and thus the number of capital units.

4.4.1 The Constant Term

Unlike the input-output coefficients which were adopted directly from the Bagchi study, the capital-output coefficients were calibrated[3]. We used capital-output coefficients from a number of studies (Kendrick and Taylor 1978, Bagchi

[3]These values are shown in Table MUK in the GAMS statement.

1984, Leontief 1973) to provide initial values for the capital-output coefficients for this study. Simulation experiments were conducted using these values and based on the model results, these initial values were then modified. If the capital-output values were not consistent with the initial values for output levels, then the solution of the model was infeasible. The calibration process continued until we were able to get: (1) a feasible solution; (2) a solution which reproduced the initial values for GDP; (3) gave a solution which was consistent with the desired paths; and (4) the capital-output coefficients were within an acceptable range in relation to those of the three studies mentioned above.

We assume that the calibrated capital-output coefficients are indicative of a period which was not environmentally aware and therefore the coefficients reflect processes which were pollution intensive. These coefficients were then used as a base from which the coefficients for the other two processes were calibrated. The pollution intermediate process was assumed to be more capital intensive than the pollution intensive and the pollution abatement technology the most capital intensive of the three.

However, the above pattern was reversed for the agriculture sector. The pollution abatement process is assumed to be land intensive with little use of capital. The pollution abatement process on the other hand is assumed to use little land but a large amount of capital. The intermediate process falls in between these two extremes. Tables 4.4 and 4.5 illustrate the capital-output coefficients calibrated for each process within the sectors in the respective regions.

Sectors	Processes		
	Pollution Abatement	Pollution Intermediate	Pollution Intensive
Industry	3.0	1.68	1.4
Agriculture	1.08	1.2	1.32
Service	2.50	1.5	1.25

Table 4.4. Capital-output coefficients for the developed region

Sectors	Processes		
	Pollution Abatement	Pollution Intermediate	Pollution Intensive
Industry	4.87	2.74	2.11
Agriculture	1.2	1.33	1.46
Service	3.59	2.15	1.65

Table 4.5. Capital-output coefficients for the developing region.

There is a substantial difference between the capital-output coefficients across the three processes in the service and industry sectors in both region. This was deliberatly done so as to emphasise the difference between the three processes in terms of costs versus CO2 emissions. The pollution abatement process decreases in capital intensiveness due to technological innovation over the planning period. However, it is assumed that the technological innovation is not sufficient to make it less capital intensive than the other two process. The difference in the coefficients do not change the basic results of the model, i.e., whether to swithch from pollution intensive to pollution intermediate or pollution abatement. However, the size of the capital-output coefficients do play a role in determing the costs incurred when swithching technologies. This is considered as one of the main reasons why the drop in GNP and consumption levels may be unrealistically high in the experiment results in Chapter Five. Nevertheless, it is the qualitative result and not the quantitative results which are of interest to us.

Furthermore, Tables 4.4 and 4.5 also illustrate that capital coefficients differ across sectors within a region as well as across regions. We assume that the industry sector is the most capital intensive, followed by the service sector and then the agriculture sector. Another assumption we make is that the developing region experiences higher capital-output coefficients than those in the developed region. This assumption was made on the belief that a major portion of the capital required by the developing region originates from the developed region and thus the extra costs of importation is reflected in the higher capital-output coefficients. Furthermore, in most cases, the technology used by the developing countries is not as efficient as those in the developed region and this is again reflected in the higher capital-output coefficients in the developing region. The computations for the other processes are shown in Appendix Two.

The capital-output coefficients are assumed to be constant throughout the planning period for the pollution-intensive and intermediate processes; i.e., technological change is not captured in the parameter calibration process. On the other hand, technological innovation is assumed to decrease the capital-intensivenes of the pollution abatement process in the industry and service sectors over the planning period. The rate of technological change is exogenously specified in the form of a one percent reduction in capital intensity per year[4]. Although this formulation does not capture the relationship between CO2 emissions and the rate of technological innovation, it serves as a good starting point to analyze the impacts of technological innovauon on policy decisions.

4.4.2 Capital-output feedback coefficients

This section deals with the computation of the feedback parameters, v_{rsp}^{k} and σ_{rsp}^{k} in equation 10 which is reproduced below.

$$\textbf{(10)} \quad \kappa_{rsp,t+1} = \mu_{rsp}^{k} + v_{rsp}^{k}\left(t_{t+1} - t_{1985}\right)$$

$$\begin{bmatrix}\text{capital output}\\ \text{coefficient}\end{bmatrix} = \begin{bmatrix}\text{fixed capital output}\\ \text{coefficient}\end{bmatrix} + \begin{bmatrix}\text{temperature rise}\\ \text{coefficient}\end{bmatrix}\begin{bmatrix}\text{change in}\\ \text{temperature}\end{bmatrix}$$

$$+ \quad \sigma_{rsp}^{k}\left(n_{t+1}^{a} - n_{1985}^{a}\right) \qquad r \in R, \; s \in S, \; p \in P, \; t \in T$$

$$+\begin{bmatrix}\text{CO2 change}\\ \text{coefficient}\end{bmatrix}\begin{bmatrix}\text{change in CO2}\\ \text{concentrations}\end{bmatrix}$$

There is no consistent and reliable data available which can be used to estimate the effects of temperature and atmospheric CO2 concentration changes on agricultural output. However, various agencies like the Central Intelligence Agency (CIA) and the United Nations Environment Programme (UNEP) have done simulation studies to analyze these effects with crop models. For example, UNEP has estimated that a doubling of atmospheric CO2 concentration levels will increase wheat and rice yields by approximately 30 and 10 percent ceteris

---

[4]This is shown in the parameter TECK in the GAMS statement.

paribus all other conditions, i.e., water and temperature. But as atmospheric CO2 concentration increases, temperatures increase which then affect precipitation patterns. Scientists postulate that the expected increase in yields from the CO2 fertilization effect may be dominated by the negative effects caused by temperature and precipitation changes.

Therefore, due to the high degree of uncertainty surrounding the feedback effects, we approached this problem by relying on sensitivity analysis using plausible ranges for the feedback coefficient estimates. This methodology will enable us to identify worst and best case scenarios.

4.4.2.1 Calibration of feedback effects of increased CO2 concentration levels

First, the effect of atmospheric CO2 concentration changes on agricultural output were computed. The underlying assumption behind this coefficient is that as atmospheric CO2 concentration increases, plant yields increase due to the CO2 fertilization effect. The results predicted by most agronomists are based on the effects on plant yields caused by a doubling of atmospheric CO2 concentration levels. Therefore, based on UNEP estimates which state that wheat and rice yields increase by 30 and 10 percent if CO2 concentrations double, we calculated the effect of a unit increase in CO2 concentration. The simplifying assumption adopted here is the linear relationship between yields and CO2 concentration levels. Although a logarithmic function which illustrates diminishing yields with increasing atmospheric CO2 concentration levels would have been more realistic, it was decided that a linear function would ensure computational simplicity and at the same time capture the relationship between yields and CO2 concentrations to a reasonable approximation.

A doubling of CO2 concentration from a steady state level of 270 parts per million volume, i.e., 540 ppmv, would cause wheat yields to increase by 30 percent. We then reasoned that if increased CO2 levels can increase yields by 30 percent, we could produce the present amount of yield with less capital. The procedure below for the developed region illustrates the methodology used to compute the σ^{k}_{rsp} coefficient:

1). From the constant term in Table 4.4 we know that in the case of the pollution-intensive process, it takes 1.32 units of capital to produce a unit of output. Now with a doubling of CO2, yields increase by 30%. Therefore, 1.32 units of capital now produces 1.3 units of output.

2) Therefore, in the new scenario, it would take only 1.02 (1.32/1.3) units of capital to produce one unit of output. The change in the capital-output coefficient is 0.3 units. This is for an increase of 270 ppmv in atmospheric concentration.

3) Therefore, for an increase of one ppmv, the capital difference is equal to 0.3/270 which gives 0.001. But as an increase in CO2 concentration increases yield and thus requires less capital, we assign negative signs to the coefficients.

4) This process is repeated for the other two processes in the agriculture sector.

Regions	σ^k_{rsp} Coefficients for the processes		
	Pollution Abatement	Pollution Intermediate	Pollution Intensive
Developed	-0.0009	-0.001	-0.001
Developing	-0.0004	-0.0004	-0.0005

Table 4.6 σ^k_{rsp} coefficients for CO2 rise effect on capital-output productivity.

The feedback effects of a climate change are assumed to affect only the agriculture sector but not the industry and service sectors.

4.4.2.2 Calibration of feedback effects of increased temperatures

Unlike the rise in atmospheric CO2 concentrations levels, a rise in temperature is expected to decrease agricultural output (Bach 1984 pg 169). The decrease is represented by the v^k_{rsp} coefficient. There are a number of reasons which can cause the decrease; the primary reason is caused by heat stress on the plants followed closely by the change in precipitation patterns. However, the response is not uniform across all crops and regions. The intensity of heat stress varies among different crops while changes in precipitation patterns differs across regions.

The primary source of data we use to calibrate the v^k_{rsp} coefficient is adopted from a study conducted by the Central Intelligence Agency (CIA 1978 pg 15). In this study, various scenarios were created for different regions around the world in which the possible effects of a climate change on agricultural output were tabulated. We use their results but with some simplifying assumptions to make the calibrations simpler. The CIA study had formulated a matrix in which the effects of temperature and precipitation changes were tabulated against the expected yields as is shown in Figure 4.1.

+10°C

			-97%	-90%	-87%	-90%	-97%		
		-95%	-77%	-54%	-43%	-54%	-77%	-95%	
	-97%	-79%	-43%	0%	14%	0%	-43%	-77%	-97%
	-95%	-70%	-17%	25%	37%	25%	-17%	-70%	-95%
	-97%	-77%	-43%	0%	14%	0%	-43%	-77%	-97%
		-95%	-77%	-54%	-43%	-54%	-77%	-95%	
			-97%	-90%	-87%	-90%	-97%		

change in temperature

-10°C

-100% 0% +100%

change in precipitation

Figure 4.1 Percentage change in crop yield for temperature and precipitation changes

From the matrix shown in Figure 4.1, we can then estimate the effect of a climate change in agriculture output. In the original CIA study, a bivariate normal distribution was associated with the matrix of yields to enable agronomists to forecast expected crop yields based on the probabilities of temperature and precipitation change. However, in this study, we assume that the temperature change caused by increased levels of atmospheric CO2 concentrations will occur and the amount by which the temperature rises is known with full certainty.

We ignore the effect of precipitation changes in this study. Scientists are still unsure of the change in rainfall patterns caused by global warming for the different parts of the globe. Furthermore, the low degree of spatial resolution used in this study is inadequate to reflect precipitation changes in a complementary manner. Nevertheless, for the moment we use the estimates given in Figure 4.1 and use sensitivity analysis to investigate the policy responses as these feedback variables are varied over a plausible range. We expect these estimates to improve as agronomists increase their knowledge on crop responses to climate changes.

Therefore, based on the simplifying assumptions above, the square in Figure 4.1 which denotes a temperature change of 2 degrees but with no change in precipitation is chosen to illustrate the effect of a temperature change on agriculture output for both regions. We assume that the effect of temperature rise on crop yields is the same for both regions as knowledge on how different crops react to heat stress is still at an early stage. The v_{rsp}^{k} coefficient for the pollution intensive process in the developed region shown in Table 4.7 was calibrated as follows:

1) The effect of a one degree rise in temperature is expected to decrease yields by approximately 25%.

2) This means that at present level of capital-output coefficients, output can be maintained at only a 75% rate.

3) The increase in capital required to maintain output constant is then equal to 1.76 (1.32/0.75), an increase of 0.44 or 33%.

Regions	V_{rsp}^{k} Coefficients for the processes		
	Pollution Abatement	Pollution Intermediate	Pollution Intensive
Developed	0.36	0.4	0.44
Developing	0.4	0.44	0.48

Table 4.7 V_{rsp}^{k} coefficients for temperature rise effect on capital-output productivity.

Although the feedback coefficients have some simplifying assumptions underlying their calibration, the formulation is sufficient to reflect the effects of a climate change on the agricultural sector. Furthermore, keeping in line with the IPCC Working Group Two's observation, we have not reflected the effect of climate change on drop in output but have captured it in the effect on the inputs for the production process. Therefore, to maintain agricultural output at desired levels, higher capital and land inputs would have to be diverted from other sectors. This formulation thus implicitly captures the negative effects of climate change on the whole economic system.

4.5 Land-Output Coefficients

Similar to the capital-output coefficient, the land-output coefficients for the production functions are also dependent on three factors: a constant term, μ^d_{rsp}, which illustrates the level of capital required to produce a single unit of output in lieu of environmental feedbacks; a second term, v^d_{rsp}, which captures the effect of temperature changes on output level and thus the number of capital units; and a third term, σ^d_{rsp}. which reflects the CO2 fertilization effect on agricultural output and thus the number of capital units. Equation 11 below incorporates the three factors mentioned above.

$$\textbf{(11)} \quad \pi_{rsp,t+1} = \mu^d_{rsp} + v^d_{rsp}\left(t_{t+1} - t_{1985}\right)$$

$$\begin{bmatrix}\text{land output}\\ \text{coefficient}\\ \text{in period } t+1\end{bmatrix} = \begin{bmatrix}\text{fixed land}\\ \text{output}\\ \text{coefficient}\end{bmatrix} + \begin{bmatrix}\text{temperature}\\ \text{rise}\\ \text{coefficient}\end{bmatrix}\begin{bmatrix}\text{change in}\\ \text{temperature from}\\ \text{base year}\end{bmatrix}$$

$$+ \quad \sigma^d_{rsp}\left(n^a_{t+1} - n^a_{1985}\right) \qquad r \in R \quad s \in S \quad p \in P \quad t \in T$$

$$+\begin{bmatrix}\text{CO2 change}\\ \text{coefficient}\end{bmatrix}\begin{bmatrix}\text{change in atmospheric}\\ \text{CO2 concentrations}\\ \text{from base year}\end{bmatrix}$$

4.5.1 The Constant Term

Although land is an important input for production processes, it is considered as part of capital in most economic models. Therefore, land-output coefficients are a bit more difficult to compute. Unlike the constant term in the capital-output coefficient which was adopted from Bagchi's study, the land-output coefficients were calibrated from data compiled in the Food and Agriculture Organization 1985 yearbook (FAO.1985,Vol 39, pg 58). To begin with, agricultural output levels for each region are tabulated. The figures are shown in Table 4.14[5].

Regions	Sectors		
	Agriculture	Industry	Service
Developed	2420	6900	8440
Developing	1200	1415	1300

Table 4.14 Sectorial output levels in billion of 1980 dollars

The next step is to compile the total land used for agricultural production in the respective regions. Table 4.8 shows the amount of arable land which was used for agricultural purposes in 1985 (ibid).

Region	Arable Land Available
Developed	676
Developing	800

Table 4.8. Land available in million hectares for regions in 1985

Once these two values have been compiled, a simple division gives the output per unit of land. However, as there is no data available which indicates the

[5]Refer to section 4.6 to see how output levels are computed from GDP figures.

type of technology which is used to grow the crops, a number of assumptions are made to facilitate the calculation of the land-output coefficients for the three different processes. We call the result the base computation.

We assume that the three processes differ in the proportion of capital and land which is used. The developing region's base computation for the land-output coefficient is assumed to be the land-intensive process, i.e., pollution abatement technology. On the other hand, the developed region's base computation is assumed to be indicative of the capital-intensive process. The land-output coefficients for the other two processes are then calculated based on base computation result[6].

Therefore, based on the arable land available and the output levels in the agricultural sector, the land-output coefficients for the respective regions are shown in Tables 4.9 and 4.10. The details of the computations are outlined in Appendix Two.

Sectors	μ^{d}_{rsp} for the Processes		
	Pollution Abatement	Pollution Intermediate	Pollution Intensive
Industry	0.0	0.0	0.0
Agriculture	1.0	0.67	0.28
Service	0.0	0.0	0.0

Table 4.9 Land-output coefficients for the developed region

[6]These coefficients are illustrated in Table MUD in the GAMS statement.

Sectors	μ^{d}_{rsp} for the Processes		
	Pollution Abatement	Pollution Intermediate	Pollution Intensive
Industry	0.00	0.00	0.00
Agriculture	0.95	0.67	0.28
Service	0.0	0.0	0.0

Table 4.10 Land-output coefficients for the developing region in million hectares per billion 1980 dollars.

As Tables 4.9 and 4.10 illustrate, only the agriculture sector uses land as an input factor for production. We have assumed that agriculture is the major consumer of land and the other two sectors use a negligible amount of land at the present time. However, as populations increase, this assumption may not be appropriate and may have to be modified so as to reflect land-output coefficients for the other two sectors too.

The major difference between the two regions is in the abatement process. The developing region's base land-output coefficient is delineated by the pollution abatement process and we assume that this is the present state of the agriculture sector. The developed region on the other hand has the comparative advantage in the pollution intensive process which incidentally is also the capital intensive process.

Unlike the other two sectors, the processes in the agricultural sector do not emit CO2. The primary emissions from agricultural practices is not CO2 but nitrogen dioxide and methane. But as this study focuses on CO2 emissions, these two gases have been ignored. Therefore, the only difference between the processes in the agriculture sector is the combination of land and capital each process uses. For notional simplicity, we call the process which is land intensive the abatement process and the process which is capital intensive the pollution intensive process. This distinction will be appropriate when the other GHG gases are included in the future; for example, higher level of fertilizer use (capital) would imply a higher emission level of the greenhouse gas nitrous oxide.

4.5.2 Feedback effects caused by perturbations to the climate system

As in the case of the capital, the feedback effects caused by changes in global surface temperatures and atmospheric CO2 concentration levels will affect output levels in the agricultural sector. But as output levels change, the amount of land units which are required to produce unit output will also be affected. We use the same methodology as in the case of the capital-output coefficients to capture these changes in the land-output coefficient.

4.5.2 1 Feedback effects from CO2 rises on the land-output variable

The effect of a doubling of CO2 levels is estimated to increase yields by 30 percent. Therefore, if all else is kept constant, the amount of land needed to maintain output at pre-climate change is estimated to decrease from 0.6 to 0.46 units of land. The 0.14 decrease is for a rise of 270 ppmv; therefore, for a rise of 1 ppmv, land input can be decreased by 0.0005 units. These coefficients are given a negative sign to reflect their positive effect on land productivity. Table 4.11 shows the coefficients for all three processes for both the regions.

Regions	σ^d_{rsp} for the Processes		
	Pollution Abatement	Pollution Intermediate	Pollution Intensive
Developed	-0.0008	-0.0005	-0.0002
Developing	-0.0003	-0.0002	-0.0001

Table 4.11 σ^d_{rsp} coefficients for CO2 rise effect on land-output productivity.

4.5.2.2 The effect of temperature rise on land-output coefficient

In the case of temperature, a two degree rise of temperature decreases yield by 50 percent; for a unit increase, output drops by approximately 25 percent. Then, if all else is kept constant, the number of land units required to maintain output levels increases from 0.6 to 0.8, i.e., an increase of 0.2 units per degree rise in temperature. Table 4.12 shows the v^d_{rsp} coefficients for the land-output coefficient.

Regions	v^d_{rsp} for the Processes		
	Pollution Abatement	Pollution Intermediate	Pollution Intensive
Developed	0.33	0.22	0.09
Developing	0.32	0.22	0.09

Table 4.12 v^d_{rsp} coefficients for temperature rise effect on land-output productivity.

Before we leave the feedback parameters, we should like to emphasis that these parameters have a high degree of uncertainty surrounding them. Therefore, in Chapter 6, we do a series of sensitivity experiments by varying these coefficients over a plausible range of values.

4.6 Energy-Output Coefficients

The energy output coefficients indicate the amount of energy required to produce a unit of output. A number of sources were used to derive these energy-output coefficients.

First, GDP for the U.S.A was obtained from the United Nations software database called Macroeconomic Data base (MEDS). This was approximately 3.1 trillion 1980 dollars. World GDP in 1985 was then approximated to be 12.4 trillion dollars since the U.S GDP accounts for 25 percent of world GDP. Then using data compiled by the World Bank, (World Development Report 1985, Table 3, pg 178) we computed the GDP share of the developed countries to be approximately 80 percent of world GDP i.e., 10 trillion dollars. The remaining 2.4 trillion is accounted for by the group of developing countries. The next step was to identify the sectorial GDP level within each region. The World Development Report 1985 (ibid) by the World Bank was used to identify the approximate proportion each sector in each region contributed to total GDP[7]. Table 4.13 shows the approximate share of each sector's GDP as a percentage of regional GDP.

[7]The manufacturing and construction sectors were aggregated as the industry sector

Regions	Sectors		
	Agriculture	Industry	Service
Developed	10%	35%	55%
Developing	35%	36%	29%

Table 4.13. Sectoral share of regional GDP

Once the sectorial GDP has been tabulated, sectorial output can be computed. Equation 1 below illustrates the relationship between sectorial GDP and output via the input-output coefficients in matrix A[8].

(1)
$$z = x - Ax$$
$$z = (I - A)x$$
$$x = (I - A)^{-1} z$$

where:

z	=	sectorial GDP
x	=	Sectorial output
A	=	Input-output matrix

Equation 1 describes a set of 3 simultaneous equations with three unknowns for each region. The solution of these equations is shown in Table 4.14.

[8]The A matrix used here is the same A matrix presented in section 4.2.

Regions	Sectors		
	Agriculture	Industry	Service
Developed	2420	6900	8440
Developing	1200	1415	1300

Table 4.14. Sectorial output levels in billion of 1980 dollars

The next step is to calculate the total amount of energy units consumed by the world in 1985. Data from the U.N. 1985 Energy Statistics Yearbook(Table 3, pg 59) indicates that the total amount of energy units consumed by the world in 1895 was approximately 254 exajoules[9]. Of the 254 exajoules, 186 exajoules (73%) were consumed by the developed[10] countries while 68 exajoules (27%) were consumed by the developing countries[11].

---

[9]Energy units supplied by electricity are not counted as this may cause double counting. Therefore, only energy units from coal, oil, and natural gas are added to get total energy consumed. Furthermore, the proportion of energy supplied by electricity is negligible compared to the three fossil fuels.

[10]Developed consists of North America, Europe, USSR, and Australia.

[11]The developing region consists of Africa, South America, and Asia.

The primary sources of these energy units were coal, oil, and natural gas[12]. Table 4.15 shows the breakdown of the 254 exajoules provided by coal, oil, and natural gas.

	Fuel Types		
	Coal	Oil	Natural Gas
Exajoules	89	105	60
% share	35%	41%	24%

Table 4.15 World consumption of energy in exajoules

From Table 4.15 it can be inferred that oil is the major source of world energy followed closely by coal and then natural gas. However, this pattern is not indicative of the fuel use patterns in the respective regions. Table 4.16 shows the proportion of energy provided by the respective fuels at the regional level.

Region	Fuel Types		
	Coal	Oil	Natural Gas
Developed	57(31%)	76(41%)	53(28%)
Developing	32(47%)	29(43%)	7(10%)

Table 4.16. Regional consumption of fuel types. (United Nations 1985 Energy Statistics Yearbook, Table 3, pg 58-85).

---

[12]Energy provided by electricity was not included to prevent double counting. Furthermore, the amount of energy supplied by electricity was too minimal to make any difference in the results.

It can also be observed from Table 4.16 that the developed region obtains its major share of energy from oil followed by coal and then natural gas. However, in the case of the developing region, coal is the largest provider of energy followed by oil. Natural gas does not play an integral part in the developing region's energy demands.

The final figures which are of interest are the proportions of world supply of the three fossil fuels that each region uses. Table 4.17[13] gives the breakdown of fossil fuel consumption by the two regions.

Region	Fuel Types		
	Coal	Oil	Natural Gas
Developed	57(64%)	76(72%)	53(88%)
Developing	32(36%)	29(28%)	7(12%)

Table 4.17. Regional share of world fossil fuel supply

In other words, our calculations indicate that the developed region uses approximately 64 percent of the world's supply of coal, 72 percent of oil, and 88 percent of natural gas. In contrast, the developing region uses 36 percent of the world's supply of coal, 28 percent of oil, and 12 percent of natural gas.

We deduce from Table 4.17 that the developed region plays a dominant role in the consumption of fossil fuels. Another interesting observation from Tables 4.17 and 4.16 is that although coal is only the second largest provider of energy for the developed region, the amount still exceeds the amount of coal consumed

[13]The figures are computed from Table s 4.15 and 4.16.

by the developing region; coal is the primary source of energy in the developing region. This observation underscores the relative disparity of fossil fuel use by the two regions.

Once the amount of energy units provided by the three fossil fuels have been accounted, the next step is to differentiate the amount of energy consumed at the sectorial level. The approximations we make to calibrate these figures were based on a study by Tirpak and Lashoff (1990, pg 161). Although, in their study, they identified three major sectors as the primary consumers of energy, we re-categorized these sectors to ensure compatibility and consistency with the sector classification in this study. Therefore, the original three sectors of transport, commercial/residential, and manufacturing were reclassified as following: the transport sector was categorized as part of the service sector while the other two were classified as part of the industry sector.

In accordance with their study results, the agricultural sector is assumed to be energy independent while the other two sectors are considered to be energy driven. We partitioned the consumption of energy between these two sectors by assigning 60 percent of total energy use to the industry sector and the remaining 40 percent to the service sector for the developed region. In the case of the developing region, the industry sector is assumed to use 70 percent while the service sector uses 30 percent. Based on these assumptions on the pattern of energy use, Table 4.18 shows the amount of energy each sector would have consumed in 1985.

Region	Sectors		
	Agriculture	Industry	Service
Developed	0.0	111.6	74.4
Developing	0.0	47.6	20.4

Table 4.18. Energy consumption in exajoules by sectors in respective regions.

The last step to get the energy-output coefficients is then to use the results in Tables 4.14 and 4.18. Table 4.14 gives sectorial output while Table 4.18 shows the sectorial energy consumption. Therefore, if energy consumed is divided by output, we get the energy-output coefficients shown in Table 4.19.

Region	Sectors		
	Agriculture	Industry	Service
Developed	0.0	0.016	0.009
Developing	0.0	0.034	0.016

Table 4.19 Energy-output coefficient in Exajoules per billion dollars; Table ZetaCO2 in GAMS.

4.7 Carbon Dioxide Emission Coefficients

The CO_2 emission coefficients are based on the physical properties of the respective fossil fuels. Table 4.20 gives the emission figures for the three fossil fuels in teragrams per exajoules (Edmonds 1986, pg 22)[14].

Fuel Type	Emission factor
Coal	23.8
Oil	19.2
Gas	13.7

Table 4.20. CO_2 emission coefficients in teragrams per exajoules

Coal emits the largest amount of CO_2 per energy unit followed by oil and then natural gas.

There are three processes in the model with each process representing technologies driven by energy supplied by the three fossil fuels mentioned above as well as renewable energy sources. The pollution abatement process is assumed to have zero level CO_2 emission as are the renewable energy sources like solar, thermal, and hydroelectric. The pollution intermediate process is assumed to be representative of a process which obtains energy from oil and natural gas. The third process, pollution intensive is associated with technology primarily driven by coal produced energy. Therefore, depending on the amount of energy provided by these processes, the amount of CO_2 emitted can be tabulated with the emission coefficients represented in Table 4.20. However, in the case of the pollution intermediate process which reflects a combination of oil and natural gas, a simple

[14]A simple weighted emission figure was computed for oil and natural gas.

average [(19.2+13.7)/2] of the emission coefficients for each fuel is used to estimate the amount of CO_2 emitted.

The next step is to convert the emission figures in Table 4.20 from teragrams per exajoules to gigatons per exajoules. This is easily done by converting teragrams to kilograms and then using the conversion figure of 2.2 lbs per kg to get the emission figures in lbs per exajoules. This figure is then divided by $2240*10^9$ to get the gigaton equivalent. Table 4.20a gives the emission figures in gigatons per exajoules.

Sectors	Processes		
	Pollution Abatement	Pollution Intermediate	Pollution Intensive
Industry	0.0	0.0177	0.0238
Agriculture	0.0	0.0	0.0
Service	0.0	0.0177	0.0238

Table4.20a CO_2 emissions in gigatons per exajoules.

4.8 Carbon Dioxide Emission from Deforestation

Deforestation is another source of anthropogenic carbon dioxide emissions.The rate of CO_2 emission differs according to the type of forests cut and burned; tropical rainforests emit larger amounts of CO_2 per hectare than temperate forests. However, present knowledge on CO_2 emission coefficients from deforestation is not as precise as those for the fossil fuels. Therefore, a degree of approximation is adopted to derive these coefficients. First, the amount of CO_2 emitted into the atmosphere by deforestation is computed. This is done by adding the amount of CO_2 emitted by fossil fuels. Then with the present knowledge of the carbon cycle, we can estimate the CO_2 emissions and absorption by the natural reservoirs. The difference between the two sources mentioned above and the empirical observations of the increase in atmospheric CO_2 concentrations would give the approximate amount emitted by deforestation. Although the figures are not precise, the values are assumed to be close enough to capture the significance of deforestation activities on the global carbon cycle.

Once the amount of CO_2 emitted by deforestation has been estimated, the next step is to approximate the level of deforestation in terms of acreage. The final step is then to calculate the emission figures by dividing the total CO_2 emitted by deforestation activities by the area of forest cut down. This gives emission factors in terms of tons of CO_2 per hectare.

Since the main source of present deforestation is in the tropics, i.e., the developing countries, the emission figure was calculated for tropical forests. As a major part of the developed region consists of countries which have predominantly temperate forests, the emission figure for the developed region was downgraded by 50 percent from those of the tropical forests to reflect emission of CO_2 by temperate forests. Although this figure has a high degree of uncertainty, it is expected that deforestation in the developed countries will not play a major role in future CO_2 emissions. Therefore, the emission figures for the developed region were included for the sake of completeness rather than for any crucial role. The calculations used to derive the CO_2 emission factors shown in Table 4.21 from are given below:

1) Approximate acreage of tropical forests cut down per year is 11 million hectares (World Resources 1988, pg 171)

2) Approximate CO2 emission emitted by deforestation activities is 1.4Gt. (IPCC Scientific Assessment 1990, pg 13).

3) Therefore, CO2 emission per hectare of tropical forest is 1.4 Gt of carbon divided by 11 million hectares which gives 0.12 Gt per million hectare.

Regions	CO2 emission coefficients
Developed	0.06
Developing	0.12

Table 4.21. CO2 emission coefficients for deforestation in the respective regions.

4.9 Initial Conditions

In general, initial values need to be specified for variables whose dynamic properties are represented by difference or differential equations. Initial conditions provide a starting point for the solution of a dynamic model. In the holistic model, initial conditions for capital stock, land stock, CO2 concentration levels in the three reservoirs, sectorial GDP levels, and investment levels were set. However, of the five variables, only the first three are crucial for the solution process. The last two are required to set the desired growth projections for these two variables in the objective function.

4.9.1 Capital stocks

Capital stocks for 1985 (base year) were calculated from the output levels and the capital-output coefficients. The initial level of capital stocks was set in a manner to reflect the combination of processes which are presently in use. We assume that the proportion of output produced from pollution abatement processes in the industry and service sectors is nil. The amounts produced by the other two processes were partitioned accordingly to the proportion of coal, oil, and natural gas used in the regions. Therefore, in the case of the developed region which derived 65 percent of its energy demand from a combination of oil and natural gas (United Nations Energy Statistics Yearbook 1985, p 58-85) and 35 % from coal, the capital stocks in 1985 were calibrated to reflect this combination.

4.9.1.1 Capital stocks in industry sector in developed region

1) 65 percent of output in the industry sector amounts to 4485 billion dollars. 35 percent of output in the industry sector amounts to 2415 billion dollars.

2) 4485 billion dollars produced by pollution intermediate process requires 7535 billion dollars of capital. 2415 billion dollars produced by pollution intensive process requires 3381 billion dollars of capital.

3) Total capital in industry sector is therefore equal to 10,916 billion dollars.

4.9.1.2 Capital Stocks in Service Sector in Developed Region

1) 65 percent of total output in service sector is 5486 billion dollars. 65 percent produced by pollution intermediate process requires 8229 billion dollars of capital stocks.

2) 35 percent of total output in service sector is 2954 billion dollars. 35 percent produced by pollution intensive process requires 3693 billion dollars of capital stocks.

3) Total amount of capital stock in service sector in the developed region is equal to 11922 billion dollars worth of capital stocks.

4.9.1.3 Capital Stocks in Agriculture Sector in Developed Region

The agricultural sector in the developed region is assumed to produce its output solely by the use of pollution intensive process. Therefore, based on the sectorial output level in the agriculture sector in 1985 and the capital-output coefficient, the actual level of capital stock in the agriculture sector in the developed region in 1985 amounts to 3195 billion dollars.

4.9.1.4 Capital Stocks in Industry Sector in Developing Region

The levels of capital stocks in 1985 for the three sectors in the developing region were estimated in the same manner as in the case of the developed region. The primary difference between the two is in the proportion of output produced by the processes. Unlike the developed region, the percentage share of total output produced by the pollution intensive process is greater than output level produced by the pollution intermediate process in the industry and service sector. On the other hand, in the agriculture sector, the situation is reversed to that observed in the developed region with output solely being produced by the pollution abatement processes.

1). 53 percent of output is produced by oil and natural gas technology. 53 percent of total output of 1415 billion dollars is 750 billion dollars. The amount of capital required to produce 750 billion dollars by the pollution intermediate process amounts to 2055 billion dollars.

2) 47 percent of output is produced by coal technology. 47 percent of total output is 665 billion dollars. The amount of capital required to produce 665 billion dollars by pollution intensive process amounts to 1403 billion dollars

3) Therefore, total amount of capital stock in industry sector in 1985 is 3458 billion dollars.

4.9.1.5 Capital Stocks in Service Sector in Developing Region

1) 53 percent of output is produced by oil and natural gas technology. 53 percent of total output of 1300 billion dollars amounts to 689 billion dollars. This requires capital stocks of 1481 billion dollars.

2) 47 percent of output is produced by coal technology. 47 percent of total output is equal to 611 billion dollars. This requires capital stocks of 1008 billion dollars.

3) Total capital stocks available in the service sector amounts to 2489 billion dollars.

4.9.1.6 Capital Stocks in Agriculture Sector in Developing Region

All output in the agriculture sector is produced by the pollution abatement process. Total output level in the agriculture sector in 1985 is estimated to be equal to 1200 billion dollars. The capital-output coefficient for the abatement process is 1.20. This therefore implies capital stocks in the agriculture sector in 1985 were in the vicinity of approximately 1440. Table 4.22 summarizes the level of capital stock available in each sector in both regions at the beginning of 1985.

Regions	Sectors		
	Agriculture	Industry	Service
Developed	3195	10916	11922
Developing	1440	3458	2489

Table 4.22. Capital stock level in 1985 (billion dollars)

4.9.2 Initial conditions for land

The initial stocks of land available in the respective regions were based on FAO's 1985 Yearbook (FAO 1985, pg 58). We compiled both arable and land under permanent crops as land available for agricultural production. Table 23 shows the initial values of land available land for production as well as forests land available for future agriculture purposes.

Type of land	**Regions**	
	Developed	Developing
Arable land	676	800
Forests	1824	2265

Table 4.23. Initial values for land (million hectares).

4.9.3 Initial values for Carbon mass in reservoirs

The carbon mass in the reservoirs is usually represented in term of gigatons carbon. Figures In Table 4.24 show the mass of carbon in each natural reservoir in 1985 (IPCC 1991, pg)

Reservoirs	Carbon Mass in 1985
Atmosphere	760
Land	2060
Surface Ocean	1005

Table 4.24 Carbon mass in reservoirs in Gt in 1985

4.10 Transfer Coefficients for the carbon cycle

The coefficients used in the carbon cycle were adapted from the carbon cycle used in the IPCC study (IPCC Scientific Assessment 1991, p 8). We assume that they are time invariant. Another assumption is that the transfer mechanism is linear over time. Table 4.25 illustrates the coefficients used in the carbon cycle[15].

Point of origin	Point of Destination		
	Atmosphere	Land	Surface Ocean
Atmosphere	N.A	0.136	0.123
Land	0.049	N.A	N.A
Surface Ocean	0.09	N.A	N.A

Table 4.25 Transfer coefficients in carbon cycle.

This brings us to the end of the data section. Since a considerable number of approximations and assumptions were made in the calibration process, to ensure consistency, the calibrated parameters were checked by running a simulation experiment called Base. In this experiment, the present world situation is simulated and the corresponding rise in atmospheric CO2 concentration and global temperature is then compared with results generated by the IPCC. If the results given by the model are consistent with those of the IPCC, then we can assume that the model's data calibration is a useful starting point for further experiments. This is the next point of discussion in chapter 5.

[15]These transfer coefficients are denoted as k in the mathematical model and are called theta in the GAMS statement.

Chapter Five

Numerical Results of Policy Experiments

5.1 Introduction

As scientific evidence substantiating the potential effects of global warming on human activities mounts, the pressure on policymakers to respond to the threat also increases. However, two factors will dictate the strategies of policymakers: (1) the concern of policymakers over the possible impacts of a CO2 emission protocol on economic growth, i.e., is it possible to reduce global CO2 emissions without compromising economic development and if not, what are the economic costs?; and (2) concern over the "right" amount of CO2 emission to cut to minimize the effects of a climate change on the economic system.

Previous studies have primarily concentrated on determining the optimal carbon tax which to reduce CO2 emissions to exogenously specified targets. However, in most of the cases, the CO2 emission constraint level was set arbitrarily at the discretion of the policymaker. These formulations have the disadvantage of not being able to derive an optimal level of CO2 emissions which minimizes the overall negative effects of CO2 emissions on the economic system, i.e., the effects of a climate change. There are two possible outcomes if the wrong level of CO2 emission cut is specified and both outcomes result in inefficient solutions to the global warming problem.

1) The first outcome arises if the level of cut is understated. In this case, the reduction in CO2 emissions will not reduce the temperature rise to a level which minimizes the feedback effects of climate change on the economic system. If this is the case, then the reduction in CO2 emissions

undertaken by the policymakers would have been in vain.

2) The second outcome arises when the level of CO_2 emission exceeds the level which minimizes the feedback effects. If this situation occurs, the cost of transition from a high CO_2 emitting economy to a low CO_2 emitting economy may have been attained at a cost higher than may have been necessary.

This chapter presents results from a number of simulation exercises which compare and contrast solutions from experiments which simulate situations similar to the present style of analysis (the setting of arbitrary CO_2 emission cuts) with those of the holistic style adopted in this study. The primary focus will be on investigating the effects on economic performance under the various regimes and to study the trade-off issues which arise in the dual issue of global warming and economic development. In the process of analyzing the results from the experiments, issues pertaining to technology substitution within the regions as well as transfer of pollution abatement technology will be addressed.

It is important to realize that the results derived from the model should not be interpreted as predictions for economic and environmental variables for the next 50 years but as pointers to: (1) important cause-effect relationships between economic and environmental variables; (2) the role of dynamics in these economic-ecology-economic relationships; and (3) the impacts of present policies on these relationships over an extended period of time. Therefore, the results should be viewed as responses to "what if" questions.

Two factors will increase the analytical power of the model. The first factor is related to the degree of uncertainty related to the anthropogenic greenhouse effect. As scientific knowledge of the global warming phenomena and the feedback effects of the warming on the economic system improves, the forecasting qualities of the model will improve too. The second factor which limits the extent of model functionality is the absence of consistent data sets for global modeling. Global economic modeling is still in its infancy; therefore, consistent data sets are difficult to find. The process of aggregating high spatial resolution data to a lower degree of resolution requires large institutional participation and is beyond an individual's capacity, but the entrance of international institutions like the World Bank and the United Nations will make global modeling a more precise endeavor.

A total of five experiments were conducted. The first four are conducted with a model structure without the feedback effects while the last experiment is conducted with a model structure with the feedback effects. The primary reason for conducting the first four experiments without the feedback loop while the last experiment has the loop is to demonstrate to the reader the importance of the feedback relationships as well as to highlight how erroneous results can be obtained if these feedback loops are not incorporated within the model structure. We should like to point out that a majority of present generation global warming models conduct experiments which fall into the category represented by the first four experiments. As mentioned before, one of the goals of this study is to illustrate the importance of the feedback loops between climate change and the economic sector. We believe that by comparing the results between experiments with and without the feedback loops would help us achieve this goal.

The first experiment is called BASE and its objective is to simulate the present trend in economic indicators as well as to predict the rate of carbon dioxide emissions over the next fifty years. The next three experiments are extensions of the BASE and depict the present style used by policymakers to study the trade-offs between economic performance and carbon dioxide emissions. The objective of these three experiments is to demonstrate the incompleteness of the present style of analyzing the trade-offs between economic performance and environmental degradation. The fifth and final experiment puts forward the holistic model structure developed in this study. The goal of this experiment is to point out the completeness of the holistic approach in conducting an "analysis of trade-offs" between economic performance and environmental degradation.

Each section will begin with a discussion on the underlying assumptions adopted for the experiment, followed by a brief explanation on the mechanics by which the objectives of the experiment are achieved. Once the basic premises of the experiments have been set, the most significant results pertaining to the experiment are presented.

(1) The primary objective of the first experiment was to simulate the present trend in the regions and to project the level of CO2 emissions for the next fifty years based on this trend. The feedback effects of climate change are ignored in this experiment. This experiment is called "Business As Usual Experiment" (BASE).

(2) The second experiment simulates a situation in which both regions decide to cut CO2 emissions which will stabilize atmospheric CO2 concentration to present levels. The feedback effects are ignored in this experiment. The experiment is called "Stabilizing CO2 Emissions" (SCE).

(3) Experiment Three reflects a situation in which the developed region's CO2 emission levels are restricted to sustainable levels while the developing

region is not subjected to any official CO2 emission control. The feedback effects are again ignored in this experiment. Experiment three is called "PayBack" (PB).

(4) The fourth experiment is the reverse of experiment three. In this scenario, the developing region is required to control its CO2 emission to sustainable levels while the developed region has no CO2 emission level imposed on it. The aid links associated with CO2 cuts by the developing region are incorporated into the model structure but the feedback equations are still ignored. This simulation run is called "Bribe" (B).

(5) The fifth and final experiment is a repeat of experiment one but with the feedback effects included in the model. This experiment is called "Holistic Base" (HB). Results from this experiment will serve as the common denominator for comparing results from the last three experiments.

Before we begin with the experiments, the first step is to set a desired rate for economic growth which policymakers expect over the next 50 years. This is achieved by setting exogenous growth rates for the state and control variables in the quadratic tracking function which is used as the objective function in this study. The priority each region places on each variable is then controlled by the weights in the priority matrix.

Table 5.1 shows the desired growth rates each region expects for the respective sectors over the fifty year planning period.

VARIABLES	SECTORS		
	Agriculture	Industry	Service
GDP	3%	3%	3%
Consumption	0.5%	3%	3%
Investment	3%	3%	3%

Table 5.1. Desired growth rates for state and control variables over a fifty year time period for the developed region.

VARIABLES	SECTORS		
	Agriculture	Industry	Service
GDP	4%	8%	2%
Consumption	4%	8%	2%
Investment	3%	4%	2%

Table 5.2. Desired growth rates for state and control variables over a fifty year time period for the developing region.

The growth rates for overall GDP in each region are shown in Table 5.3. The figures are computed from the sectorial growth rates.

INDICATOR	REGION	
	Developed	Developing
GDP	3%	6%
Consumption	3%	5.4%
Investment	2.8%	3.5%

Table 5.3 Regional growth rates for economic variables.

Table 5.4 below shows the desired levels for CO2 emissions and deforestation rates for both regions. The desired level of CO2 emissions was based on the scientific premise that global CO2 emissions need to be reduced by 60 percent of present levels if atmospheric CO2 concentration is to be stabilized at the present level of 354 ppmv. Therefore, world economic output is allowed to emit 3 Gt of carbon equivalent. This amount is then divided equally between the two regions. Although this may not be a realistic division, this value can be easily changed in future experiments to simulate various policies, e.g., CO2 emission per capita. The deforestation rates chosen for the study are similar to present trends observed in the regions (World Resources 1988, pg 171). We did not set a lower rate of deforestation for the developing region because we believed that these countries will not cut deforestation rates if no incentives are given by the developed countries. As the model does not capture aid tied to deforestation levels, we believed that it would be more appropriate to follow present trends.

Variables	Regions	
	Developed	**Developing**
Deforestation	1 million hectare per year	11 million hectares per year
CO2 emission	1.5 Gt carbon per year	1.5 Gt carbon per year

Table 5.4 Desired path for environmental variables for both regions.

5.2 Experiment One: Business As Usual (BASE)

The primary objective of this experiment is to test the consistency of the model. This is done by simulating present economic trends in the regions and then extrapolating the future levels of atmospheric CO2 concentration and temperature rises based on these trends. If the results are in agreement with those of the Intergovernmental Panel on Climate Change (IPCC), the model validity is to a certain extent established. In addition to testing the validity of the model, the BASE simulation is also used as a learning experiment to provide insights to crucial relationships which "drive" the solution procedure in the economic model.

5.2.1 Experimental Assumptions

The following assumptions are used in the Business As Usual experiment:

1) The first assumption specifies that both regions emphasize economic growth without any deep commitment to curbing CO2 emissions. This assumption simulates present trends observed in the regions.

2) The second assumption enables the model to simulate the present use of technological process. It is achieved by placing a lower bound on the level of output produced by the use of pollution intermediate processes. If a restriction is not placed on the use of processes, the optimal solution will reflect a process combination which is the least capital intensive; in this case the pollution intensive process which derives its energy supply from the burning of coal. However, the present mix of technology in the regions reflects an equal division between coal on one hand and oil and natural gas on the other hand. Therefore, to capture this process mix, a lower bound is placed on the amount of output that is produced by pollution intermediate process (energy from oil and natural gas).

5.2.2 Experimental Mechanics

The weights on the CO2 emission variable for both regions are set at low levels relative to the weights on all other variables. This means that tracking the desired path for CO2 emission by the two regions is not an important issue; in other words, there is no constraint on CO2 emissions by either region.

To be consistent with the present policy making process, only the economic sub-model is used in the BASE experiment; the feedback loop between climate change and the economic system is ignored.

5.2.3 Experimental Results

The absence of a commitment to the maintenance of a sustainable level of CO2 emissions allows both regions to choose a combination of pollution intensive and intermediate techniques in the energy dependent sectors. This combination primarily arises because of the lower bound on pollution intermediate output levels. In the absence of this requirement, the combination of processes would in all probability have primarily consisted of pollution intensive processes. However, the primary objective of the experiment is to capture the predominant use of coal, oil and natural gas and the results in the base solution reflect this present state of technology mix in the regions to a close approximation. Figure 5.1 compares the CO2 emission levels observed in the BASE solution to those of two projections put forward by the IPCC.

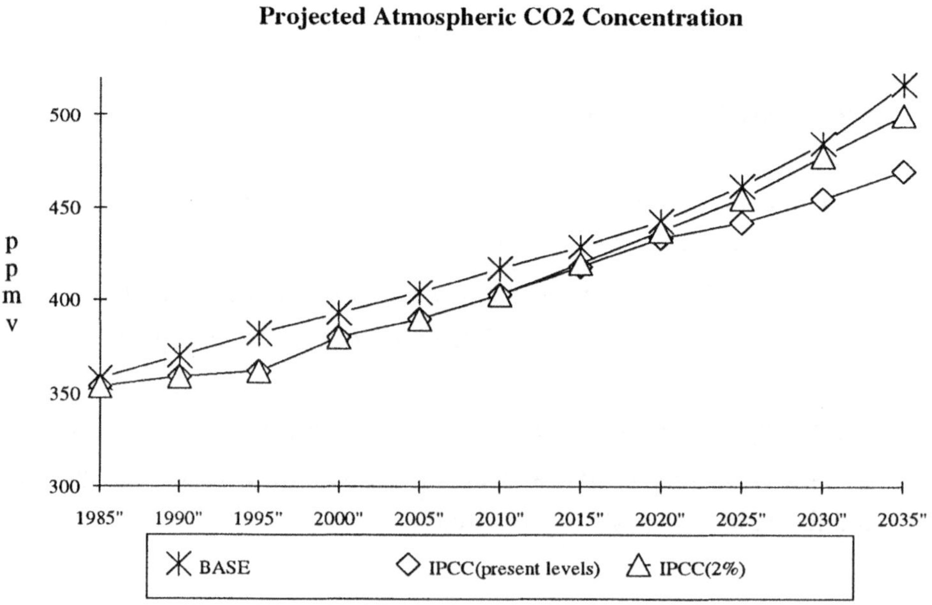

Figure 5.1 Projected global CO2 emissions.

Based on the technology combination used by both regions in the Base experiment, the level of CO2 emissions is expected to increase significantly over the next fifty years. The projections exceed those of the IPCC's high emission scenario (IPCC 2% 1990 pg 15) by approximately five percent. Taking into account of the large degree of uncertainty associated with predicting future levels of CO2 emissions and the corresponding increase in atmospheric CO2 concentrations, this difference is considered negligible.

The question which arises next is: Who is responsible for the increase in atmospheric CO2 concentrations? Are the developed countries still the primary emitters or will the developing countries surpass them? In response to the question, Figure 5.2 below shows the developing region overtaking the developed region in total CO2 emissions by the year 2015.

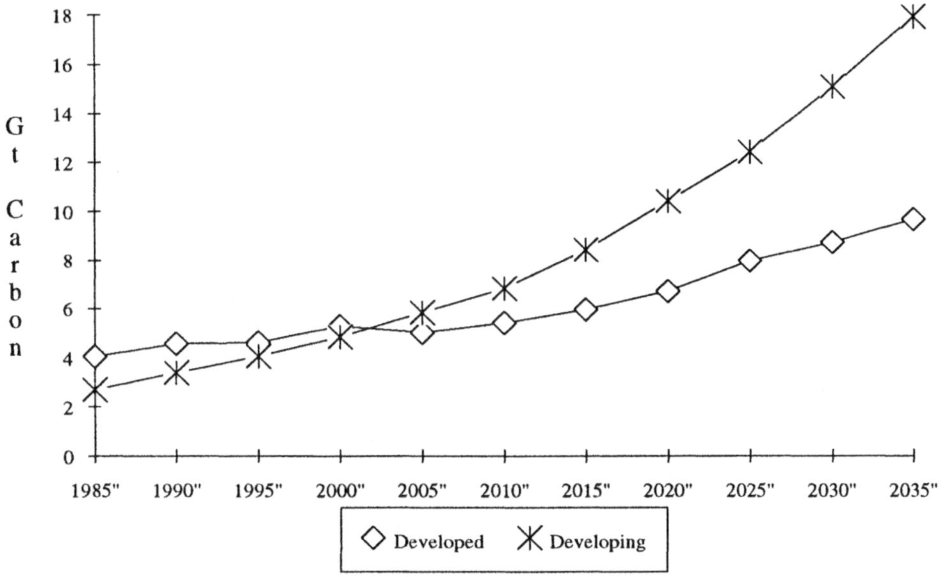

Figure 5.2 CO2 emissions by the respective regions.

The primary factor contributing to the developing region's increase in CO_2 emissions is the increased use in fossil fuels. The contribution from deforestation also increases over time but the share of its contribution to total emission decreases from 65 percent in 1985 to approximately 15 percent by the end of the planning period. However, this result is dictated to a certain extent by the upper bound on the deforestation rate. Therefore, if the developing region is allowed to increase its rate of deforestation, the share of total CO_2 emission attributed to deforestation will increase thus increasing CO_2 emissions by the developing country.

The economic results from the BASE experiment highlight some important constraints which determine the overall economic growth of both regions over the fifty year planning period. The land constraint in both regions plays an integral part in determining the final output in all three sectors. The analysis is as follows. First, the level of output in the agriculture sector is determined by the amount of capital and land available for production i.e., the factor inputs to the production process. However, of the two inputs, the land constraint proves to be more stringent and ultimately determines the level of output in the agriculture sector in both regions.

The output levels in the other two sectors do not face a factor constraint as capital accumulation grows at a sufficient rate to achieve desired levels[1]; but as agricultural output is used as an intermediate good for the production process of these two sectors, the final output level in both these two sectors is constrained to a certain extent by the output level in the agriculture sector. Therefore, the final result reflects a compromise between reduced consumption levels in the agriculture sector and a larger proportion used as an intermediate good by the other two sectors.

---

[1]The industry and services sector do not use land as an input for production.

INDICATOR	REGION	
	Developed	Developing
GDP	2.7%	5.5%
Consumption	2.8%	5.4%
Investment	2.3%	5.7%

Table 5.5 Regional growth rates for economic variables in the Base experiment

Even if the figures in Table 5.5 indicate that the overall regional economic performance of both regions do not deviate too much from the desired levels, Figures 5.3 and 5.4 indicate significant differences between the desired and BASE GDP levels in the agriculture sector for both regions.

The discrepancy is hidden at the regional level due to the relatively small size of the agriculture sector in the developed region. The zero level of GDP arises primarily from the low priority placed on agriculture output in the last few time periods. Although the weights have not been changed, the size of the sectors have changed considerably with respect to each other and therefore if the same priorities are to be maintained throughout the planning period, the weights have to be changed with time. As this was not done in the study, the priority placed on agriculture GDP fell with time. We did not attempt to change the weights accordingly with time in this study as the emphasis was on analyzing the trade-offs between carbon emissions and economic performance under various regimes. Therefore, if the weights were changed, the effect would have been replicated for all experiments and the basic results would not have changed.

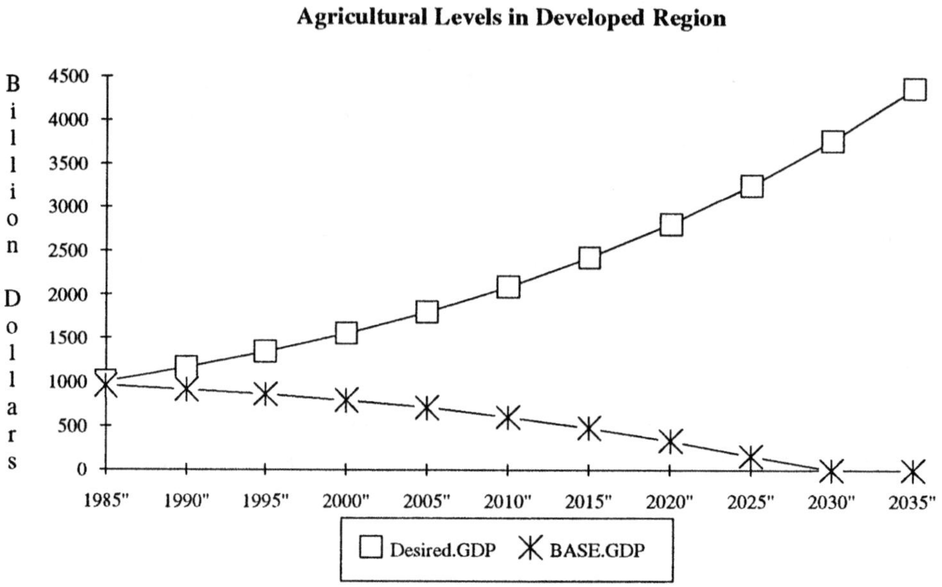

Figure 5.3 Agriculture GDP levels in the developed region

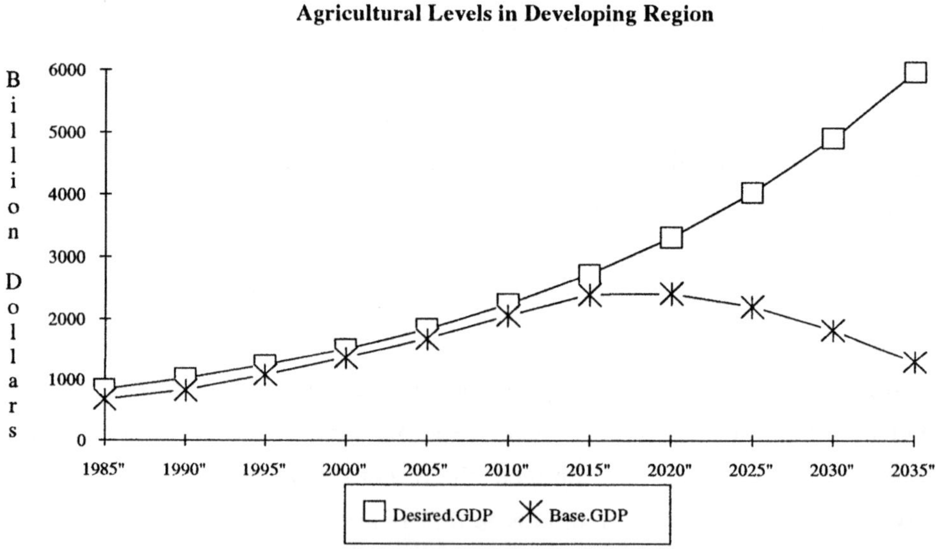

Figure 5.4 Agriculture GDP in the developing region.

In the case of the developing region, although the agriculture sector accounts for a significant share of the economy in the early part of the period, its size diminishes as the region industrializes and the other two sectors grow rapidly.

Unlike the energy dependent sectors in which both regions choose a combination of pollution intensive and intermediate technology, the agriculture sector in the developing region chooses a combination of pollution abatement and intermediate processes while the developed region used primarily the pollution intensive technique. The difference arises from each region's comparative advantage in a particular technology.

The developing region with its comparative advantage in land intensive techniques chooses the pollution abatement process[2] in the early stages of the planning period. But as the land constraint becomes tight in the later part of the planning period, the region shifts from pollution abatement techniques to the intermediate processes which are less land intensive. On the other hand, the developed region uses its comparative advantage in capital intensive techniques in the agriculture sector and thus opts to use the pollution intensive process throughout the planning period.

Figures 5.5 and 5.6 illustrate that although agricultural output in both regions increase over time the GDP level decreases.

---

[2]From Table 4.1 to 4.3, it can be inferred that the developing and the underdeveloped regions have a comparative advantage in the land intensive pollution abatement processes whilst the developed region has a comparative advantage in the capital intensive pollution intensive technique.

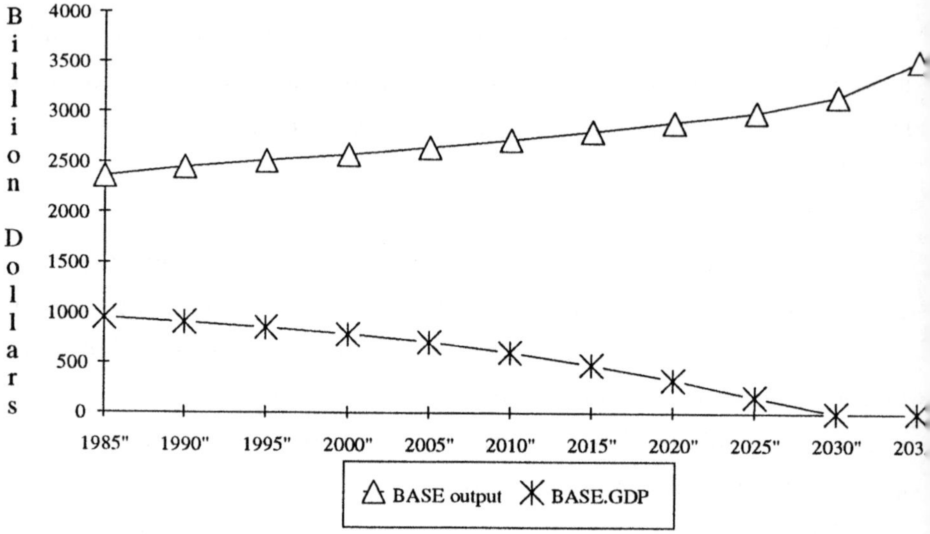

Figure 5.5 Output and GDP levels in the developed region.

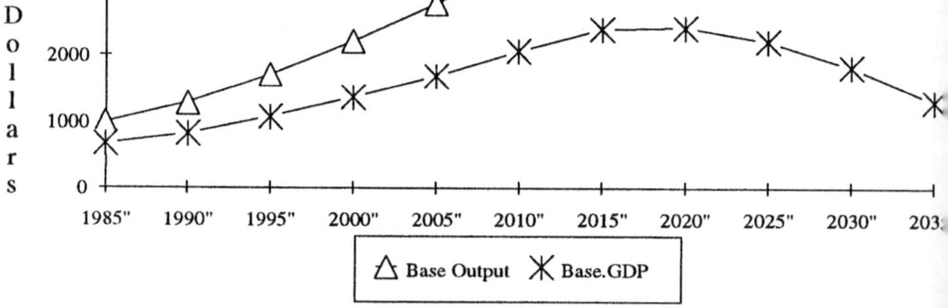

Figure 5.6 Output and GDP levels in the developing region.

This is primarily caused by the larger absorption of agricultural output by the other two sectors as intermediate goods. Therefore, if GDP levels in the agricultural sector in both regions are to increase, output has to grow at a larger rate. However, the land constraint in both regions places an upper bound on the output level. Therefore, the only way output can be increased is by either: (1) a structural change in the input-output coefficients to reflect a smaller share of agricultural outputs being used as intermediate goods; or (2) an increase in technological innovation in the agricultural sector to cause an increase in land productivity.

Previous studies have tended to classify land under the general category of capital, but as populations increase and the pressure for larger supplies of agricultural output increase, the land availability issue becomes an important factor. The significant effect of land availability for future agricultural production to feed a growing population makes it imperative that further research be conducted in this area.

5.3 Experiment Two: Stabilizing CO2 Emission Levels (SCE)

There has been considerable discussion among policymakers as to the level of carbon dioxide emissions which needs to be cut. Scientists assert that if atmospheric carbon dioxide concentrations are to be stabilized at existing levels, then carbon dioxide emissions need to reduced by 60 percent of present levels; approximately 3 to 3.5 gt of carbon per year. We assume that policymakers accept this figure and therefore enforce a policy stating that total carbon dioxide emissions in terms of carbon is 3 Gt a year. In order to simplify the analysis, the 3 Gt of carbon emission is partitioned equally between the two regions, i.e., each region is allowed to emit 1.5 Gt of carbon a year. A potential area of future research is to experiment with partitions which are more equitable, for example, a division based on population levels in the region.

In order to enforce an official CO2 emission protocol, the weights on the CO2 emission variable in the priority matrix of the criterion function were increased significantly[3] with respect to the economic variables. In other words, each region wants the deviation between the optimal level and the desired level of the CO2 emission to be zero; in this case the desired path for CO2 emission is set to be approximately 1.5 Gt carbon per year for each region.

5.3.1 Experiment Results

With an official policy calling for a global cut in CO2 emissions, the rise in atmospheric CO2 concentration was observed to have risen to 385 ppmv by the year 2035. This rise in atmospheric carbon dioxide is approximately 79 percent lower than the level observed in BASE. As expected, the temperature rise was significantly lower under SCE conditions as compared to BASE; 0.3 degree Celsius rise versus 1.6 degree Celsius rise respectively.

The next step is to investigate the process by which carbon dioxide emissions were reduced and to analyze the impact the process had on economic performance for the two regions. Table 5.6 shows the economic conditions observed under the SCE strategy.

INDICATOR	REGION	
	Developed	Developing
GDP	2.6% drop	1% drop
Consumption	4.5% drop	9.7% drop
Investment	12.5% rise	14.2% rise

Table 5.6 Difference between SCE and Base results.

[3]Weights were increased from 7E-8 to 7E+8.

The results in Table 5.6 indicate that both regions experience drops in GDP and consumption levels. The primary reason for the drop in consumption levels is the presence of the official protocol on CO_2 emissions. The 1.5 Gt carbon emission constraint causes the regions to cut back drastically on consumption levels and divert resources to investment which is required to speed up the process of capital accumulation. It can also be observed from Table 5.6 that the developing region suffers a larger drop in consumption levels than that of the developed region. This can be attributed to the easier transformation process the developed region goes through when converting from pollution intensive to pollution abatement processes. The developing region, on the other hand, requires a larger amount of investment resources to transform its economy and is therefore required to reduce its consumption levels significantly.

The issue which policymakers need to resolve now is whether the costs incurred for reducing carbon dioxide emission levels by 60 percent are justified, i.e., are the drops in consumption and GDP levels matched by the benefits received by limiting the growth rate in atmospheric CO_2 concentrations and temperature rises? If the cure is worse than the illness, it may be irrational to adopt the prescription for the illness. However, this issue cannot be addressed in this model framework as the positive and negative feedback effects of climate change are not considered within the model structure. Therefore, the model structure used in this experiment does not contain sufficient information to help policymakers analyze the trade-offs between environmental degradation and economic performance when establishing CO_2 emission protocols.

5.4 Experiment Three: The Developed Region accepts a CO2 Protocol, "PayBack" (PB)

This experiment is a variation of SCE; only the developed region is subjected to a CO_2 emission protocol. The developing region is allowed to continue emitting CO_2 at present trends. This is a policy similar to the Montreal protocol which governs the emission of CFC's. It is called PayBack because the policy can be viewed as a situation in which the developed countries are required to pay for past emissions by accepting the burden of reducing CO2 emissions by themselves.

The weights on the developed region's CO2 emission level is increased significantly with respect to all other variables[4]. This has the function of thus reducing CO2 emissions by the developed region to the desired level of 1.5 Gt of carbon.

5.4.1 Experimental Results

Atmospheric CO2 concentration increases from a 1985 level of 358 ppmv to 465 ppmv at the end of the planning period. The increase in atmospheric CO2 concentration causes global surface temperatures to rise by approximately 1.14 degree Celsius. In other words, a strategy calling for the developed countries to cut their CO2 emission levels by 70 percent from present levels would reduce global temperature rise by 30 percent. This point exemplifies the importance of the role that developing countries will play in the future on the global warming issue. A 70 percent cut by the developed region translates to only a 30 percent reduction in the projected temperature rise.

Table 5.7 summarizes the overall effect of the PB strategy on the economic performance of both regions over the fifty year period.

---

[4]The weight on CO2 emissions by the developed region is maintained at 7E+8 while the weight on CO2 emissions by the developing region is lowered to 7E-8.

INDICATOR	REGION	
	Developed	Developing
GDP	2.6% drop	1% drop
Consumption	4.4% drop	2.7% rise
Investment	12.5% rise	5.7% drop

Table 5.7 Difference between PB and Base for economic indicators.

The developed region experiences the same economic performance as witnessed under the SCE strategy. This is expected as the region is operating under the same conditions as in the SCE strategy. The developing region on the other hand experiences a significant increase in consumption levels in comparison with levels in the SCE experiment. Although there is no significant difference between the GDP levels under the two strategies for the developing region, the percentage share of GDP that is allocated for consumption is much higher in this experiment than in the SCE. This is because the developing region allocates less resources to capital formation: there is no pressure to invest in capital intensive pollution abatement processes due to the lack of a CO_2 emission protocol.

In conclusion, the developed region experiences again significant drops in consumption and GDP levels as witnessed in under SCE conditions. The issue of whether the costs incurred for reducing carbon dioxide emission levels i.e., the drop in consumption and GDP levels matched by the benefits received by limiting the growth rate in atmospheric CO_2 concentrations and temperature rises again? The issue still cannot be addressed as the positive and negative feedback effects of climate change are still not considered within the model structure. Therefore, the issue of whether the model structure used in this experiment adequate to help policymakers decide in establishing CO2 emission protocols rises again.

5.5 Experiment Four: The Developing Region Accepts a CO2 Protocol, the "Bribe" experiment (B).

This experiment is a reversal of experiment three; the developing region is required to cut its CO2 emission levels to 1.5 Gt of carbon per year while the developed region continues with its present trend. However, unlike experiment three, aid flows from the developed to the developing region are incorporated into the model structure. The primary reason for adding the aid flows is that it would be unrealistic to expect the developing countries to cut their CO2 emission levels without some form of incentives. This the reason the experiment is called the Bribe solution as it reflects the situation in which the developed countries are allowed to continue present emission levels by buying the emission rights of the developing countries in the form of untied aid.

As in the case of experiment three, the weights in the priority matrix were used to ensure that the developing region cut its CO2 emission levels to the desired level[5].

5.5.1 Experiment Results

The CO2 concentration increased from 358 ppmv in 1985 to 448 ppmv in 2035 which translated to a temperature rise of 0.97 degree Celsius. The increase is smaller than that observed under the PB strategy by approximately four percent. This result was expected since the Base experiment demonstrated that the developing countries exceed the developed countries in total CO2 emissions over the planning period.

---

[5]The weight on CO2 emissions in the developing region was increased from 7E-8 to 7E+8 and the weights on the developed region's emissions was reduced from 7E+8 to 7E-8.

Table 5.8 below summarizes the economic results observed under the B strategy.

INDICATOR	REGION	
	Developed	**Developing**
GDP	1.3% drop	2.2% drop
Consumption	7.2% drop	7.6% rise
Investment	12.5% drop	17.7% rise

Table 5.8 Difference between B and BASE for economic indicators

As Table 5.8 indicates, the developed region experiences a lower level of economic performance. This is primarily caused by the presence of the financial aid flows. In the absence of these aid incentives, the developed region would have experienced an economic performance similar to the BASE experiment. However, with the added burden of providing aid to the developing region, the developed region has to divert resources which otherwise would have been used for domestic investment and consumption. This diversion of resources explains the lower levels of these variables in the developed region. An interesting observation arises when the results from Table 5.7 and 5.8 are compared. The developed region, if given the choice, would choose the PB strategy over the B strategy even if it meant adopting a CO2 protocol; it experiences higher levels of domestic consumption and investment under the PB strategy.

On the other hand, the B strategy is an attractive option for the developing region. Although it experiences a drop in GDP levels, the foreign aid flows enable it to increase its consumption and investment levels considerably. The drop in GDP arises primarily from the increased demand in capital stock which is necessary for the transformation process from a pollution intensive to a pollution abatement economy. The aid from the developed region is diverted to consumption and investment purposes but the share partitioned to investment is not sufficiently large enough to increase capital accumulation to a level necessary to increase GDP beyond the Base levels.

However, the issue of whether the costs incurred for reducing carbon dioxide emission levels i.e., the drop in consumption and GDP levels in the developed region, are matched by the benefits accrued to the region by limiting the growth rate in atmospheric CO2 concentrations and temperature arises. Again, the model framework does not give a full representation of all the factors for a complete "analysis of trade-offs" to be conducted. The major lacking factor in this experiment as well as in the last two experiments is the feedback effects of a climate change on the economic system. As long as these feedback effects are not considered in the decision process, a complete analysis on the trade-offs between economic performance and environmental degradation cannot be conducted. Furthermore, the main point of contention in this and the previous two experiments is between economic performance and the cost of reducing carbon dioxide emission levels. However, it is not the cost and level of carbon dioxide emissions which is the important factor but more so the magnitude of the feedback effects caused by the climate change which is then inadvertently determined by the level of carbon dioxide emissions. Therefore, it is the trade-off between the change in economic performance and the feedback effects of climate change, a cost/benefit analysis and not just a cost analysis, should be the focus of debate. The fifth and final experiment is designed to address this issue.

5.6 Experiment Five: Holistic Model Simulation (Holistic Base)

The fifth and final experiment approaches the issue of economic performance and global warming from a totally different perspective as compared to the last three experiments. In this experiment, the economic model used in the previous experiments is expanded to include the feedback effects between climate change and economic productivity. The main focus of this experiment is to investigate how present policy decisions will be affected if policymakers take into account the feedback effects of global warming into the policy decision making process, i.e., a holistic approach. Ecological economists assert that the protection of environmental integrity is a necessary condition for sustained economic development in the long run. The holistic model formulation style adopted in this experiment will enable us to test this hypothesis. The assumptions adopted for this experiment are identical to those in the BASE experiment one with the exception that the lower bound on pollution intermediate processes is relaxed.

5.6.1 Experiment Results

The key result from this experiment as compared to the BASE solution is the decrease in the rate of growth of atmospheric CO2 concentration. As there is no official policy requiring both regions to curb CO2 emissions, CO2 emissions would have been expected to follow the trend witnessed in the BASE experiment. However, global CO2 emissions were only 49 percent of those in the BASE experiment. Therefore, by the inclusion of the feedback effects to the model structure, CO2 emissions were reduced by 51 percent over the fifty year period. The CO2 emission cut translated to a temperature rise of 0.64 degree Celsius as compared with the 1.6 degree Celsius rise observed in the BASE experiment.

Both regions reduced their respective emission levels with the developed region experiencing a 85 percent cut in emissions while the developing region cut its CO2 emissions by 20 percent. As figure 5.7 shows, the developing region emits a larger amount of CO2 than the developed region throughout the planning period, beginning from 1990.

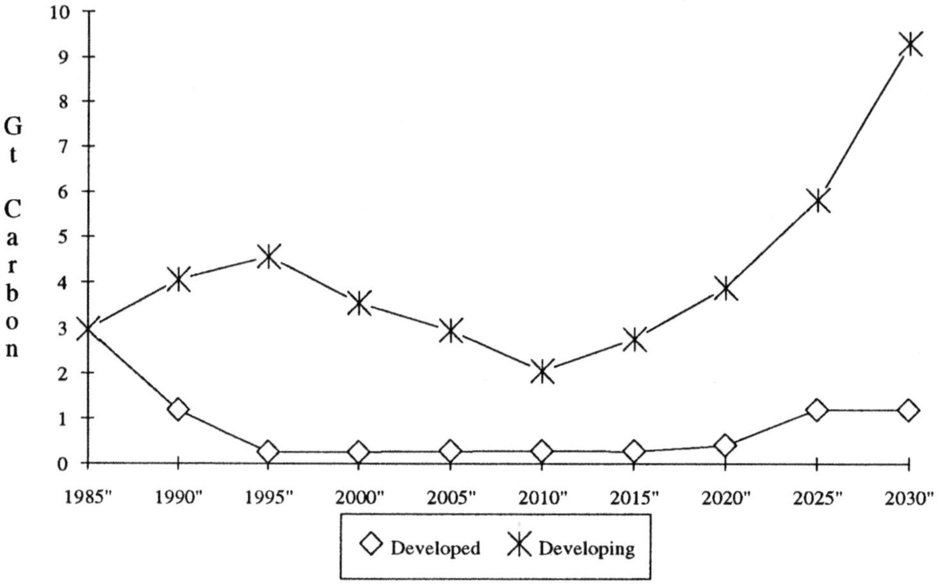

Figure 5.7 CO2 emissions by the two regions under Holistic Base conditions.

The difference in CO2 reductions by the two regions is due to the fact that the developed region accounts for the larger share of CO2 emission in the first part of the planning period; and it is more efficient to minimize increases in atmospheric CO2 concentration by cutting CO2 emissions early in the planning period (IPCC 1990). The model formulation takes advantage of this factor and thus takes the appropriate steps to cut CO2 emissions early which then translates to cuts in CO2 emissions by the developed region.

The observation in the last paragraph raises the question: What mechanisms are used in the model to cut CO2 emissions? The main impetus which is causing a lower level of CO2 emissions comes from the switch from pollution intermediate and intensive processes to a technology combination consisting primarily of pollution abatement processes. The driving force behind the choice of processes is the feedback loop between environmental change and economic productivity. The feedback effect causes a significant drop in economic productivity, in particular the productivity of land in the agriculture sector. Therefore, to minimize the drop in productivity, CO2 emissions are reduced; and the primary mechanism used in the model to minimize the feedback effects is through the substitution of high CO2 emitting processes with low CO2 emitting processes.

As a result of switching from pollution intensive to pollution abatement processes, both regions experience a drop in GDP and consumption levels. However, investment increased in both regions. Table 5.9 shows the percentage difference between the BASE experiment and this experiment.

INDICATOR	**REGION**	
	Developed	**Developing**
GDP	4.8% drop	1.8% drop
Consumption	7.4% drop	5.5% drop
Investment	14.6% increase	7.7% increase

Table 5.9 Difference between BASE and Holistic Base for economic indicators.

The reason behind these results is the switch from pollution intensive processes to pollution abatement processes in the industry and service sectors. Unlike the BASE experiment, the optimal combination of processes used by both regions in this experiment consisted primarily of pollution abatement technology and to a lesser extent the pollution intermediate processes. However, the rate of transformation differed significantly across the two regions. The developed region adopted the technology combination consisting primarily of pollution abatement processes throughout the planning period beginning from the base year. The developing region, on the other hand, adopted a more gradual approach to the technology switch by having a portion of its total output produced by pollution intensive processes in the first part of the planning period. This difference in behavior can again be attributed to the time preference for cutting CO2 emissions early in the planning period, thus placing higher pressure on the developed region to switch at a faster rate as it is the larger emitter of CO2 emissions in the first part of the planning period.

The Holistic Base experiment has demonstrated that if policymakers take into account the feedback effects of a climate change on economic productivity, there is an implicit mechanism which causes policymakers to switch to pollution abatement processes early in the planning period. This mechanism by itself reduces CO2 emissions by approximately 50 percent from the "Business as Usual" experiment. In other words, a 50 percent reduction in CO2 emissions is required to minimize the effect of climate change on economic productivity.

The magnitude of the feedback effects is reflected in the κ_{rspt} (capital-output) and π_{rspt} (land-output) coefficients. If there is no climate change, ceteris paribus, these two variables are constant and do not change over time. However, if there is a change in atmospheric CO2 concentrations and global temperatures, the κ_{rspt} and π_{rspt} coefficients change. The final value of these coefficients is dependent on the size of the CO2 fertilization effect versus the size of the temperature effect on agricultural yields. If the former is larger than the latter, then the κ_{rspt} and π_{rspt} values decrease in value; productivity in the agriculture sector increases since the capital-output and land-output ratios are lower. On the other hand, if the temperature effect is larger than the CO2 fertilization effect, capital and land productivity in the agriculture sector decreases. One factor which could dampen the drop in productivity of the two factors is technological innovation prompted by global climate changes. We have not captured technological innovation in the present specification of κ_{rspt} and π_{rspt}. This is an appropriate subject for future research.

The results from the Holistic Base experiment point to a drop in productivity in both capital and land. As Table 5.10 indicates, the developing region experiences a larger drop in productivity than will to the developed region.

VARIABLES	REGIONS	
	Developed	**Developing**
Kappa	18.1% drop	19.8% drop
Phi	14.2% drop	17.8% drop

Table 5.10 Feedback Effects on Land and Capital Productivity.

However, the reader is reminded that the magnitude of the feedback effect is dependent on the parameter values adopted for the feedback equations. As the results in this experiment are dictated to a large extent by these parameter values, a sensitivity analysis is conducted in the next chapter to determine the lower limit to these parameter estimates which will give the same policy decisions as observed in this experiment. In other words, the sensitivity analysis will determine the minimum feedback effect which will dictate policymakers to adopt the policy decisions observed in this experiment. Therefore, if scientific evidences indicate that the feedback effects will be larger than the minimum level identified in the sensitivity analysis, then it would be wise for policymakers to embark on the type of policy decisions illustrated in the Holistic Base experiment.

We have shown in this section how economic performance and carbon dioxide emissions are related via the feedback relationships. In the following section, we shall compare the Holistic Base solution to those of SCE, PB, and B and show how the latter solutions give incomplete results if an "analysis of trade-offs" between economic development and environmental degradation is conducted. In the process, we hope to demonstrate the strengths of the holistic methodology over the existing partial approach methodologies used in the SCE, PB, and B experiments.

5.6.2 The Holistic Base compared to SCE ,PB and B strategies

On first observation, the figures in Table 5.11 will indicate that the two regions performed better under SCE conditions as compared to Holistic conditions.

INDICATOR	REGION			
	Developed		Developing	
	HB	SCE	HB	SCE
GDP	4.8% drop	2.6% drop	1.8% drop	1% drop
Consumption	7.4% drop	4.5% drop	5.5% drop	9.7% drop
Investment	14.6% rise	12.5% rise	7.7% rise	14.2% rise

Table 5.11 Results of HB and SCE with respect to BASE results.

Furthermore, the smaller rise of 0.3 degrees Celsius under SCE all point to the fact that the extent of environmental degradation is also smaller in the SCE experiment. Therefore, one could infer that the SCE strategy provided a more efficient solution than Holistic Base. However, this conclusion will hold if policymakers are only concerned with economic performance summed over the fifty year planning period. On the other hand, if policymakers are also concerned with the yearly fluctuations in economic performance, then a different scenario emerges. As Figures 5.8 and 5.9 clearly illustrate, both regions experience significant drops in consumption levels in the early part of the planning period.

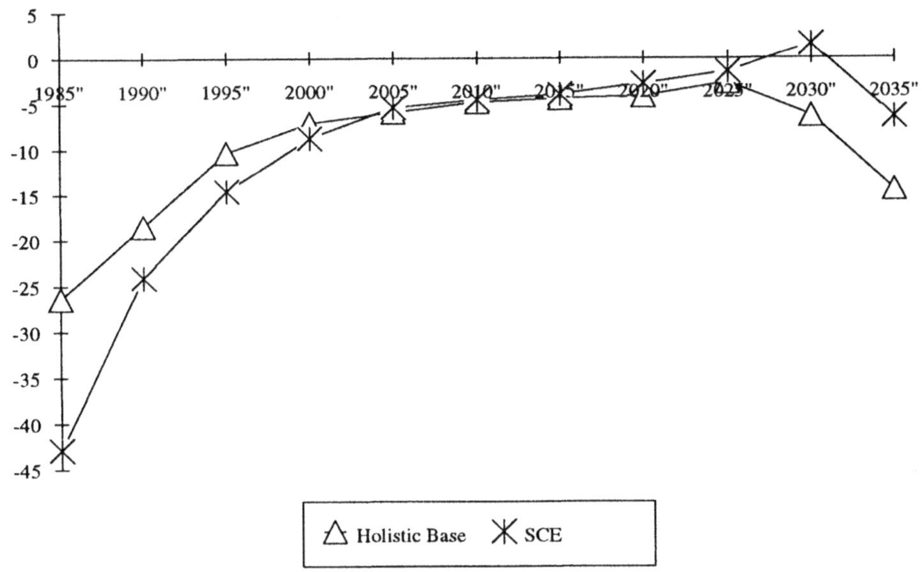

Figure 5.8 Percentage difference in consumption levels between Holistic Base and SCE experiments for the developed region.

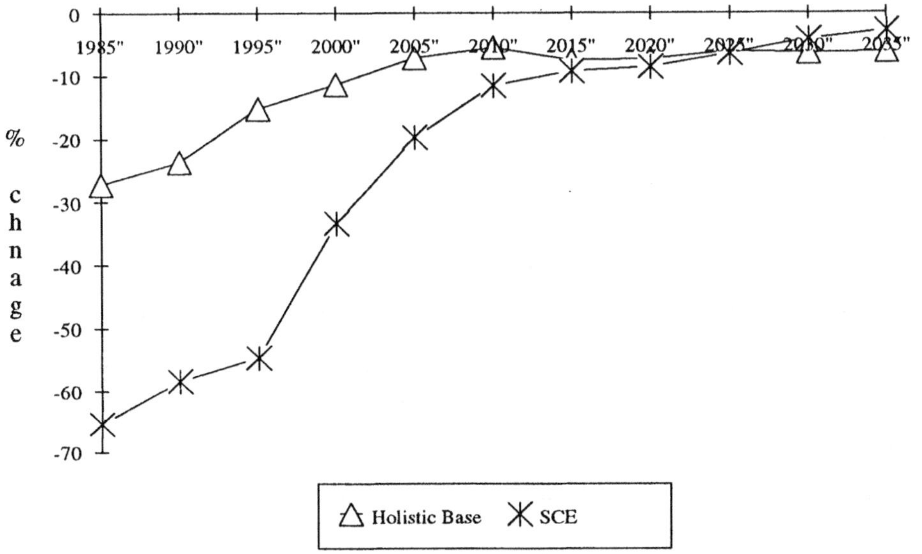

Figure 5.9 Percentage difference in consumption levels between Holistic Base and SCE experiments for the developing region

The final decision on which of the two strategies results in an overall better strategy rests on the trade-off policymakers make between: (1) smaller deviations in consumption levels in each period but with a slightly smaller overall growth rate; and (2) larger deviations in each time period but with a higher growth rate over the planning period.

We should point out that the size of drop in consumption levels in both regions is unrealistically high. A strong factor dictating the size of drop is the style the three processes in each sector have been represented. Due to the discrete relationship between the production processes, capital intensiveness, and CO2 emissions, the costs of transforming from a high to low CO2 emitting economy is dictated by lump sum rather than smooth combinations of the processes. Nevertheless, the qualitative result of the experiments do not change and thus the analytical powers of the model are not diminished.

The next level of comparison between the two strategies is from the perspective of the two regions as individual units, i.e., if given a choice, which of the two strategies will the policymakers in the two regions choose?

The developing region experiences a smaller drop in consumption levels under Holistic Base conditions while the developed region fares better under the SCE conditions. The reason for this dichotomy in results arises from the different mechanisms the two strategies use to reduce CO2 emissions. In the Holistic Base model structure, no mandatory upper bound on emission levels by each region for each year exists. On the other hand, the SCE strategy places an upper bound of 1.5 Gt on the yearly emission of carbon dioxide by each region. Figures 5.10 and 5.11 compare the rate of CO2 emission reduction under the two strategies for the two regions. Under the SCE scenario, the developed region emits a total of 15.4 Gt of carbon over the planning period; it is allowed to emit only 11.5 Gt of carbon under Holistic Base conditions.

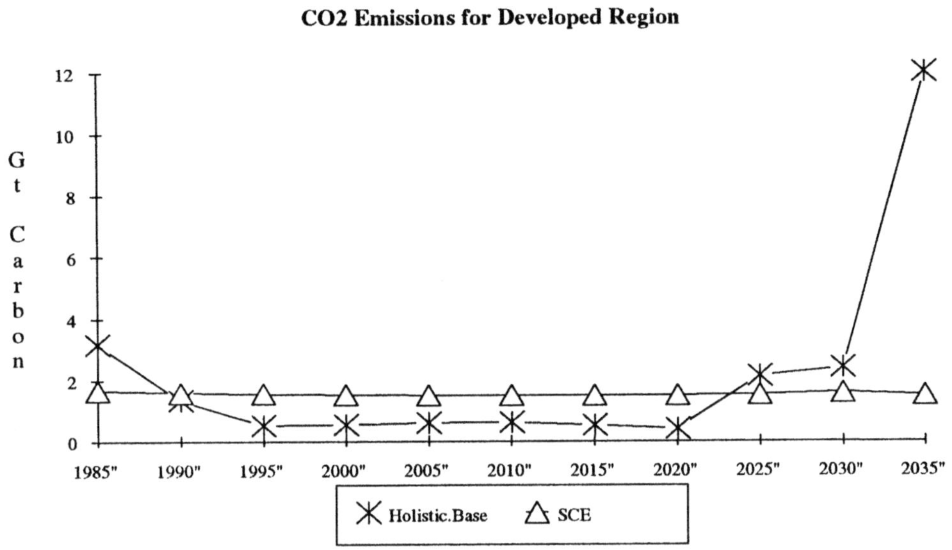

Figure 5.10 CO2 emission reduction paths for the developed region.

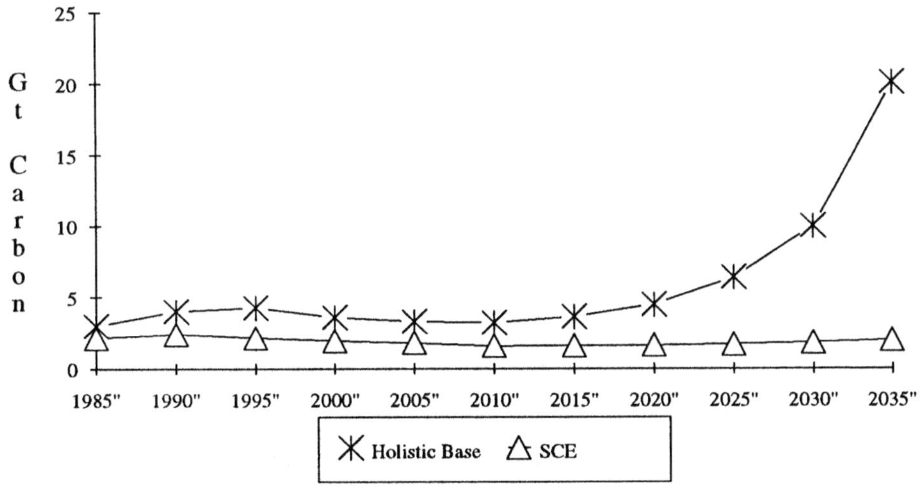

Figure 5.11 CO2 emissions for the developing region.

On the other hand, the developing region has to cut its CO2 emissions by a larger amount under the SCE strategy versus the Holistic Base. The developing region is allowed to emit 65.1 Gt of carbon under the Holistic Base scenario, but only 18.6 Gt of carbon under the SCE strategy. The developing regions fares better under Holistic conditions as the burden of reducing a major portion of future CO2 emissions falls on the developed region. This can be attributed to the "time preference" of cutting CO2 emissions in the early part of the planning period if feedback effects of global warming are to be reduced. Since the developed region is the major emitter in the early part of the planning period, it is forced to shoulder the burden of reducing CO2 emissions.

From the perspective of both regions, an international protocol on CO2 emissions will have significant consequences on the economy in the first 25 years with the developing region experiencing the largest drop in economic performance. But is the reduction and disruption witnessed under the SCE scenario necessary? The Holistic Base experiment had identified the lower bound level of temperature rise which would minimize the feedback effect of a climate change: 0.64 degree Celsius (higher than the 0.3 degree rise witnessed in the SCE strategy). Therefore, it can be inferred that the official CO2 emission reduction policy had overshot the target and had caused an unnecessary drop in economic performance. This result arises from the fact that in the SCE experiment, the benefits of averting climate change are not taken into consideration. Therefore, the "optimal" reduction level suggested by the SCE strategy results in a situation in which the costs of averting climate change exceed the benefits derived from preventing climate change. Furthermore, the SCE strategy does not provide an equitable solution for the developing region. The mere fact that both regions are subjected to the same emission protocol and that protocol is advantageous to the developed region will immediately be rejected by the developing region. Our conclusion is that the SCE policy would result in a sub-optimal solution and thus an inefficient solution.

How do the PayBack and Bribe strategies compare with the Holistic Base strategy? These two strategies, on first observation[6], appear to be superior to the Holistic Base if compared solely on economic performance. However, in both cases, the temperature rise witnessed is above the 0.64 degree Celsius threshold mark identified in Holistic Base; the lower bound before the feedback effects cause a drop in economic productivity. Therefore, 1.14 degree rise witnessed under the PB strategy indicates that the economic performance observed under this strategy is overstated; i.e., with the feedback effects, the output levels will be lower than those observed under the PB strategy. This outcome arises when the economic costs caused by drops in land and capital productivity outweigh the economic costs incurred due to the adoption of greenhouse preventative measures: this occurs when there is a temperature rise above 0.64 degrees.

In the case of the Bribe strategy, the reduction of CO2 emissions by the developing region was not sufficient to maintain the temperature rise within the 0.64 degree range; the 90 ppmv increase in atmospheric concentration causes an increase in global surface temperatures of 0.97 degrees. This indicates that economic performance reported under Bribe conditions is overstated. In reality, the temperature increases will reduce the productivity of capital and land and this will cause output levels to drop. These results occur because, similar to the SCE strategy, the benefits of preventing climate change are not considered in the analysis. The "optimal" reduction levels suggested by the PB and B strategies result in a situation in which the damage caused by the temperature rise exceeds the costs saved by adopting a limited emission reduction strategy.

[6]Compare Table 5.6 and 5.8.

5.7 Conclusion

A number of important points were highlighted by the series of experiments conducted in this chapter. The first point concerns agriculture output and the importance of land availability for future production. The Base experiment had identified land as a tight constraint which restrains agricultural output and consequently the output levels of the other sectors in the economy. The important role the agriculture sector plays makes it imperative for more research to be conducted in this area.

The second important factor to be highlighted by the experiments is the role feedback effects play on economic productivity. The results from the Holistic Base experiment suggest that these feedback effects are large enough to dictate present policy decisions. The 51 percent cut in global CO2 emissions which restrains the temperature rise to approximately 0.64 degrees suggests that this is the optimal level of CO2 reductions which is necessary to minimize the potential effects of climate change on economic productivity.

The third and key result arises from the SCE, PB, and B experiments. The practice of estimating the potential effects on the economic system by arbitrary cuts in CO2 emissions will always result in solutions which do not capture the true essence of the trade-off analysis between economic development and environmental degradation. The probability of estimating the "right" level of CO2 reduction by this method is low. The feedback effects suggest that if the temperature rise is allowed to exceed a critical point, the marginal reduction in CO2 emissions is ineffective and a waste of resources. In other words, if a CO2 emission reduction is to take place, the amount to be cut must be sufficient to minimize the feedback effects of the resulting temperature rise on the economic system. Therefore, the 32 and 43 percent reductions in CO2 emissions witnessed in the PB and B strategies would have in all probability been in vain.

These results illustrate the superior nature of the Holistic model results. The Holistic model has the advantage of conducting a marginal cost-benefit analysis within the model structure. In the case of the model structure with no feedback loops, only a cost analysis can be conducted while the benefits are ignored. The results illustrated above highlight the serious implications of this mis-specification especially when the dynamics of climate change are taken into consideration.

Chapter Six

Sensitivity Analysis

6.1 Introduction

One of the primary functions of optimization models is to highlight important cause-effect relationships inherent in a model. The series of experiments conducted in the last chapter achieved this directive by highlighting a number of important factors which dictated the results to a large extent. One of these factors was the feedback effect between climate change and economic productivity. It was concluded from the results observed under the Holistic Base experiment that the magnitude of the feedback effect of a climate change was the determining factor which caused policymakers to switch from a high CO2 emission to a low CO2 emission economy. However, this result is to a large extent dictated by the parameter values which are used in feedback equations, 10 and 11 below.

$$\textbf{(10)} \quad \kappa_{rsp,t+1} = \mu^{k}_{rsp} + v^{k}_{rsp}\left(t_{t+1} - t_{1985}\right)$$

$$\begin{bmatrix}\text{capital output}\\ \text{coefficient}\end{bmatrix} = \begin{bmatrix}\text{fixed capital output}\\ \text{coefficient}\end{bmatrix} + \begin{bmatrix}\text{temperature rise}\\ \text{coefficient}\end{bmatrix}\begin{bmatrix}\text{change in}\\ \text{temperature}\end{bmatrix}$$

$$+ \quad \sigma^{k}_{rsp}\left(n^{a}_{t+1} - n^{a}_{1985}\right) \qquad r \in R, \; s \in S, \; p \in P, \; t \in T$$

$$+\begin{bmatrix}\text{CO2 change}\\ \text{coefficient}\end{bmatrix}\begin{bmatrix}\text{change in CO2}\\ \text{concentrations}\end{bmatrix}$$

$$\textbf{(11)} \quad \pi_{rsp,t+1} = \mu^{d}_{rsp} + v^{d}_{rsp}\left(t_{t+1} - t_{1985}\right)$$

$$\begin{bmatrix}\text{land output}\\ \text{coefficient}\\ \text{in period } t+1\end{bmatrix} = \begin{bmatrix}\text{fixed land}\\ \text{output}\\ \text{coefficient}\end{bmatrix} + \begin{bmatrix}\text{temperature}\\ \text{rise}\\ \text{coefficient}\end{bmatrix}\begin{bmatrix}\text{change in}\\ \text{temperature from}\\ \text{base year}\end{bmatrix}$$

$$+ \quad \sigma^{d}_{rsp}\left(n^{a}_{t+1} - n^{a}_{1985}\right)$$

$r \in R$

$s \in S$

$p \in P$

$t \in T$

$$+\begin{bmatrix}\text{CO2 change}\\ \text{coefficient}\end{bmatrix}\begin{bmatrix}\text{change in atmospheric}\\ \text{CO2 concentrations}\\ \text{from base year}\end{bmatrix}$$

The feedback equations 10 and 11 have four parameters, v^{k}_{rsp}, σ^{k}_{rsp}, v^{d}_{rsp}, and σ^{d}_{rsp} which are computed from a variety of resources which are based on results from crop models and estimates from agronomists. These parameters were not estimated using the traditional econometric tools due to the lack of consistent data. Furthermore, projected temperature increases caused by higher levels of atmospheric CO2 concentrations have yet to be realized. Therefore, any data set that is presently available would not reflect a significant relationship between agriculture yields and temperature increases. This leaves us with the next best solution- -sensitivity analysis.

Sensitivity analysis is a useful tool for determining upper and lower bounds for important parameter values. The model is solved a number of times using varying values for the parameter under observation. A plausible range of estimates is identified when results change from one extreme to another. In the case of the climate change feedback effects, it would be beneficial for policymakers to know the range of parameter estimates which will help them to decide to either adopt the policy decisions observed in Holistic Base or Base experiments. In other words, the lower bound on the effects of a temperature rise and the upper bounds on the CO_2 fertilization effect on agriculture yields.

6.2 Sensitivity Analysis on Feedback Parameters

The parameter estimates in the Holistic Base experiment will be used as the base for comparing the results from the sensitivity analysis experiments. To maintain consistency for the comparisons, the model is simulated under the same conditions as in Holistic Base; the weighting matrix is the same as in the Holistic Base experiment and the values for all other parameters remain unchanged.

In the case of the Holistic Base experiment, the values for the feedback coefficients reflected the following assumptions:

1) A temperature rise of two degrees causes a 50 percent drop in agricultural yields in both regions.

2) A doubling of atmospheric CO_2 concentration levels causes agriculture yields to increase by 30 and 10 percent in the developed and developing regions respectively.

The first objective in the sensitivity analysis experiments is to identify the minimum effect a two degree temperature rise has on agriculture yields which would not cause policymakers to switch from a high CO_2 emission to a low CO_2 emission economy. If scientists then believe that this minimum is within an acceptable range, it would not be beneficial for policymakers to embark on any kind of CO_2 emission protocol[1].

Five experiments are conducted. The first four experiments illustrate scenarios in which the magnitude of the temperature feedback effect on agricultural yields decreases. In the fifth and last experiment, the CO_2 fertilization effect is diminished. The following assumptions govern the five experiments:

Scenario 1

1) A temperature rise of two degrees causes a 30 percent drop in agriculture yields in both regions.

2) The CO_2 fertilization effect is held at same levels as in Holistic Base.

Scenario 2

1) A temperature rise of two degrees causes a 20 percent drop in agriculture yields in both regions.

2) The CO_2 fertilization effect remains unchanged.

[1]The feedback effects in this study are limited to just the agricultural sector. A more detailed analysis will take into consideration the effects of temperature rises on other aspects of society as discussed in chapter one.

Scenario 3

1) A temperature rise of two degrees causes a 15 percent drop in agriculture yields in both regions.

2) The CO_2 fertilization effect remains unchanged.

Scenario 4

1) A temperature rise of two degrees causes a 10 percent drop in agriculture yields in both regions.

2) The CO_2 fertilization effect remains unchanged.

Scenario 5

1) A two degree rise in temperature causes a 10 percent drop in agriculture yields in both regions.

2) The CO_2 fertilization effect is reduced to 20 and 5 percent from 30 and 10 percent for the developed and developing regions respectively.

6.2.1 Experiment Results

As the intensity of the temperature feedback effect on agriculture yields is decreased, the global level of CO_2 emissions increase. With the increase in CO_2 emissions, global surface temperatures also increase. Table 6.1 below shows the levels of emissions and the corresponding temperature rises observed in the first four experiments.

SCENARIOS	VARIABLES		
	Regional CO2 Emissions in Gt Carbon		Global Temperature rise
	Developed	Developing	
Holistic Base	12.3	31.2	0.54
Scenario 1	13.7	41.4	0.88
Scenario 2	20.6	58.5	1.03
Scenario 3	38.7	62.9	1.31
Scenario 4	75.5	71.9	1.78

Table 6.1 The effects on regional CO2 emissions and global temperatures under the first four experiments.

The major contributing factor to increased levels of CO2 emissions is the technology mix used in the respective regions. As the feedback effect of temperature increase on economic productivity decreases, the impact of higher temperatures on the economic system diminishes, thus relieving pressure on the regions to reduce their level of CO2 emissions. This in turn translates into a a larger proportion of output being produced by pollution intermediate and intensive technology in both regions.

This pattern is similar for the industry and service sectors across both regions. Rather than illustrating the results of each sector in each region, either sector in either region will serve to demonstrate the major impacts of the experiments. The industry sector in the developed region is chosen to demonstrate the results from the experiments as it illustrates the transformation between the experiments more explicitly. Figure 6.1 shows the proportion of total output that was produced in the industry sector by the different processes in the developed region under the respective experiments.

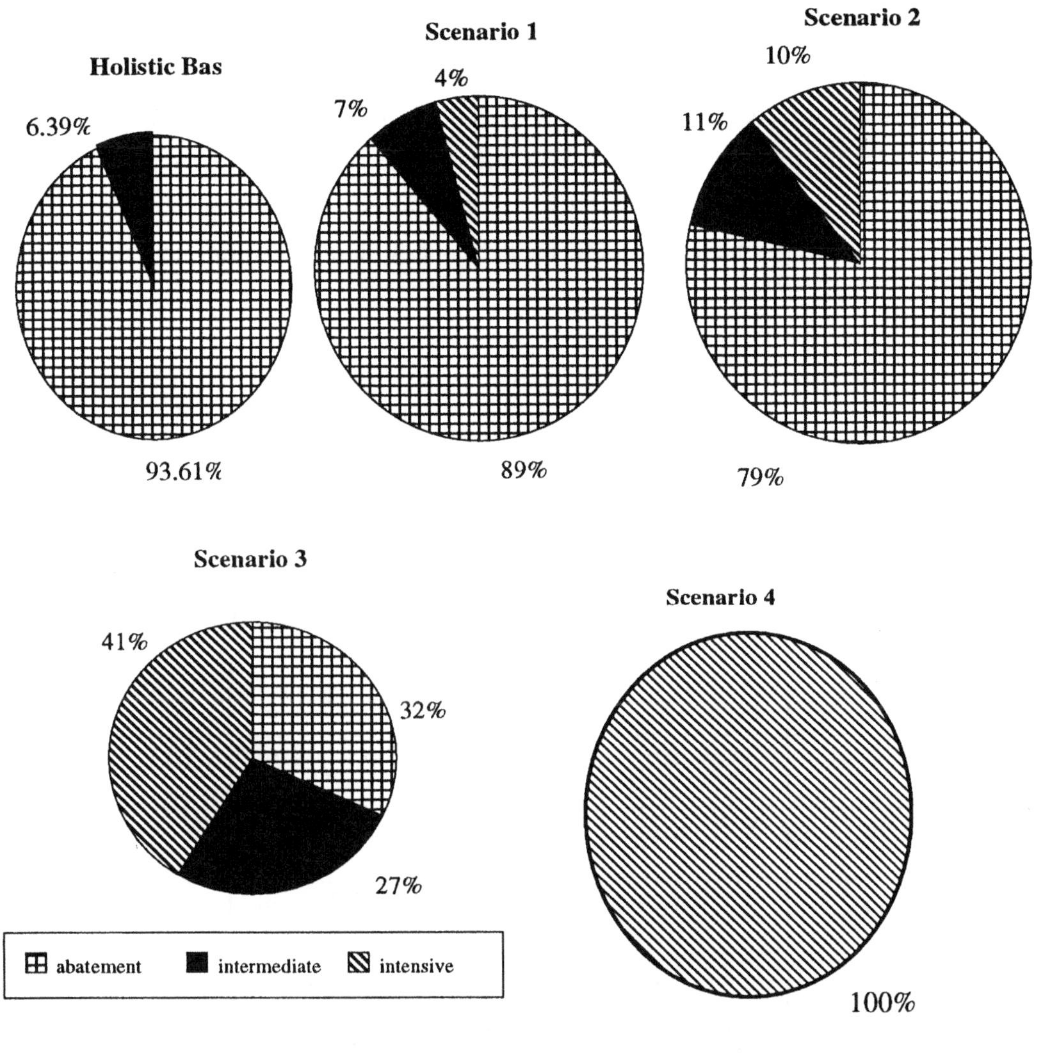

Figure 6.1 Technology mix under the different experiments.

As figure 6.1 illustrates, the use of pollution intermediate and intensive processes increases as the impact of the temperature feedback effect decreases. The turning point occurs when a two degree temperature rise is assumed to decrease agriculture yields by ten percent. At this level, the feedback effects are not significantly high enough to prompt policymakers to switch from a high CO2 not significantly high enough to prompt policymakers to switch from a high CO2 emitting to a low CO2 emitting economy. Therefore, if it is assumed that the feedback effects of a temperature rise do not exceed those observed in Scenario 4, then it can be inferred that there is no need for policymakers to respond to the global warming problem. It should be stressed at this point that this study only addresses the issue of the feedback effects of climate change on agriculture yields. A complete analysis of the problem should include all other significant feedback effects of climate change.

Table 6.2 shows the drop in economic productivity (reflected by kappa and phi variables) for both regions under the various experiments.

SCENARIOS	**REGIONS**			
	Developed		**Developing**	
	kappa	**phi**	**kappa**	**phi**
Holistic Base	9.8% drop	8.9% drop	10.9% drop	10% drop
Scenario 1	8.2% drop	7.5% drop	8% drop	9.6% drop
Scenario 2	4.5% drop	3.2% drop	6.2% drop	5.7% drop
Scenario 3	2.8% drop	1.4% drop	5.5% drop	3.6% drop
Scenario 4	1.5% rise	7.6% rise	3.4% drop	negligible

Table 6.2 Percentage change in land and capital productivity caused by temperature increase.

The results in Table 6.2[2] show that the drop in land and capital productivity decreases as the size of the temperature feedback effect is decreased. In scenario 4, the CO2 fertilization effect dominates the temperature effect and the developed

[2]The percentage changes in the land and capital productivity are computed by finding the difference between kappa and phi at the beginning and end of the planning period.

region actually experiences a rise in land and capital productivity. However, the developing region experiences a small drop in capital productivity and fares slightly better in land productivity by experiencing no change from base year levels. This ensues because the ten percent CO2 fertilization effect is not large enough to overcome the temperature effect whereas the 30 percent CO2 fertilization effect in the developed region dominates the temperature effect. It is the combination of an increase and no change in land productivity in both regions that enables policymakers to use pollution intensive technology. However, if there is a slight change in the feedback effects which may cause the land productivity in either region to drop, then in all probability, the model results will indicate a policy which requires a certain portion of the economy to adopt pollution intermediate and abatement processes.

The above inference was tested by reducing the degree of the CO2 fertilization effect on agriculture yields in both regions. The experiment is called scenario 5. Instead of a 30 and 10 percent increase in yields, a 20 and 5 percent increase is adopted for the developed and developing regions, respectively. The temperature feedback effects in scenario 5 are identical to those in scenario 4. Table 6.3 shows the effect on the capital and land productivity variables.

SCENARIOS	**REGIONS**			
	Developed		**Developing**	
	kappa	**phi**	**kappa**	**phi**
Scenario 5	0.8% drop	0.14% drop	2.7% drop	2.1% drop

Table 6.3 Results from scenario 5.

The CO2 fertilization effect is now insufficient to overcome the negative effects of temperature increases. The land productivity in the developed region has fallen slightly but the developing region experiences a larger drop in productivity. It is the effect of the fall in land productivity in the developing region which causes both regions to curb their CO2 emission levels so as to minimize the feedback effects. This result exemplifies the fact that the global warming problem is a global issue and the actions of one region can significantly effect the outcome of the other region. In this case, the larger drop in land productivity in the developing region causes both regions to adopt pollution intermediate and abatement processes. Figure 6.2 shows the technology mix adopted by the energy sector in the developed region.

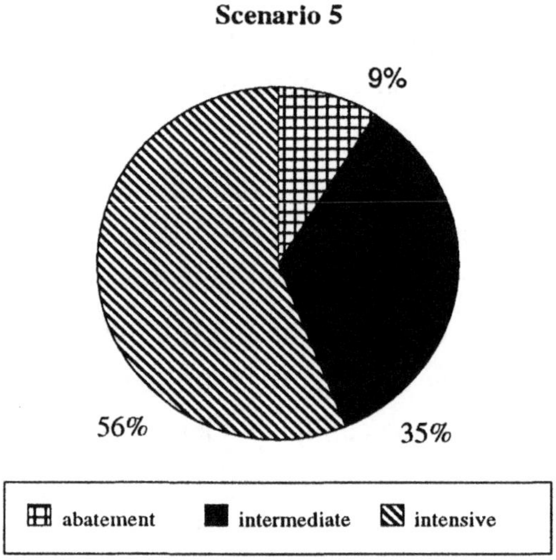

Figure 6.2 Technology mix under scenario 5 conditions.

The technology combination differs from that in scenario 4; both economies have opted to use pollution abatement and intermediate processes in this experiment. The reason for choosing the pollution abatement and pollution intermediate processes is to minimize the temperature feedback effect, achieved by curbing CO2 emissions, which in turn is achieved by adopting low CO2 emitting technology.

6.3 Sensitivity Analysis on Land-Output Coefficient

Another parameter which played a crucial role in the experiments conducted in Chapter Five was the land-output coefficient. The tight constraint on land availability dictated to a large extent the results observed in the Base and Holistic Base experiment. It was observed that the feedback effects from a climate change were sufficiently large to cause significant drops in land productivity. The land degradation caused by climate change then further exacerbated the pre-climate change land shortage problem. This was one of the main reasons underlying the decision to substitute high CO2 emission processes with low CO2 emitting technology in the Holistic Base experiment even in the absence of an official CO2 emission protocol. However, if technological innovations cause land productivity to increase over time, will the results in Holistic Base still hold or will they be reversed? In other words, will a technologically induced increase in land productivity be able to overcome the drop in land productivity caused by climate change?

We address the above question by running a series of sensitivity tests on the land-output coefficient. The constant term in the coefficient is reduced at a specified rate over the planning period. Although, this methodology has the disadvantage of not capturing the technical innovation endogenously, the exogenous specification will suffice for our needs; it can be viewed as Dis-embodied technological innovation. Four simulations were conducted. In the first experiment, land productivity was increased at 0.5 percent a year over the planning period. The rate of increase was increased to 1.0, 1.5, and 2.0 percent in the next three experiments respectively.

6.3.1 Experimental Results

As land productivity increased, the degree of temperature rise also increased. In other words, the higher the rate of technological innovation in the agriculture sector, the higher is the tolerance level for temperature rises. In the Holistic Base experiment in which there were no improvements in land productivity, the negative feedback effect of a climate change on land productivity caused policymakers in both regions to cut their CO2 emissions thus reducing the rate of drop in land productivity. In the four experiments in this section, as the rate of land productivity was increased, the level of CO2 emissions by the regions also increased. This means that the rate of drop in land productivity by climate change is negated to a certain extent by the rise in land productivity caused by technological innovations.

In fact the results indicate that the increase in land productivity arising from technological innovation exceeds the rate of drop in land productivity caused by climate change. This result has the effect of increasing output levels in the agriculture sector. It was also observed that output levels in the industry and service sectors also increase as land productivity increases. Two factors are responsible for the increase in output levels in the industry and service sectors. The first factor arises from the increased output levels in the agriculture sector which is now able to meet the intermediate needs of these two sectors. It was observed in the Base results in Chapter Five that output in the service and industry sectors was lower than desired levels because output levels in the agriculture sector were not large enough to meet both the intermediate demands by the service and industry sectors as well as for consumption purposes.

The second factor behind the increased levels in the industry and service sectors arises from the less stringent controls on CO2 emissions. As the rate of technological innovation was increased, the agriculture sector was better equipped to handle the negative feedback effects of a temperature rise. Therefore, even if higher temperatures caused large drops in land productivity, these drops were over come by the increase in technological innovation. This factor therefore relaxed some of the pressure to substitute high capital intensive abatement processes with lower capital intensive processes even if these processes emitted higher levels of CO_2. Therefore, as the agriculture sector became better equipped to handle the drop in land productivity caused by temperature increases, the objective of cutting CO_2 emissions became secondary to achieving desired output levels.

However, even with a rate of increase of 3 percent per year in land productivity over the entire planning period, the service and industry sectors in both regions still used a combination of abatement , intermediate, and intensive pollution processes. The region which cuts its CO_2 emissions the most is still the developed region. This was observed in all four experiments. This result is consistent with our earlier results, i.e., the time preference for cutting CO_2 emissions early on in the planning period to minimize temperature rises.

As mentioned in the above paragraph, even with a technology induced increase in land productivity of 3 percent per year, both regions still used a combination of all three processes. The question that arises next is, what rate of technological innovation is needed for both regions to ignore the feedback effects of temperature rises on land productivity? In other words, the rate which allows both regions to choose a technology mix comprising mainly of intermediate and intensive processes in the service and industry sectors. We therefore kept increasing the rate of land productivity until both regions chose a technology mix made up of the intermediate and intensive technology. It was observed that a rate of approximately 7 percent per year over the planning period makes policymakers to ignore the global warming issue. We can therefore conclude from these tests is that if policymakers believe that land productivity can be increased at 10 percent per year over the next 50 years, then no policy response is required to cut CO_2 emissions. However, a 7 percent productivity target may be an unrealistically high level which may not be attained. If this is the case, then the results point to a CO_2 emission reduction policy by both regions with the developed region adopting a larger cut in emissions.

6.4 Conclusion

The results from the sensitivity analysis point out that: (1) if global surface temperature increases cause a less than 10 percent drop in agriculture yields; (2) the CO_2 fertilization effect increases agriculture yields by 30 and 10 percent in the developed and developing region respectively; and (3) if increase in land productivity by technological innovation is greater than about 7 percent per year over a 50 year period, then policymakers need not respond to the global warming issue by curbing CO_2 emissions.

A key experimental result is the level of CO_2 emissions by both regions. As the emphasis on cutting CO_2 emissions decreases, the developed region experiences a larger marginal increase in CO_2 emissions. This trend is explained again by the time preference for cutting CO_2 emissions. A policy which calls for an early cut in CO_2 emissions will result in the developed region shouldering the major portion of the CO_2 emission cuts. Therefore, it is advantageous for the developing regions to pursue a holistic style of analysis rather than policies which call for an arbitrary level of emissions.

The final factor highlighted by the sensitivity analysis experiments is the importance of land for agriculture. As observed in the experiments above, the drop in land productivity causes the model the solution to include measures to minimize CO_2 emissions. It was also observed that even if one region's drop in land productivity is minimal as compared with the other's, the final result dictates that both regions cut their CO_2 emissions in order to minimize global CO_2 emissions. This result emphasizes the global nature of the greenhouse effect and the importance of international cooperation and coordination if the issue is to be solved.

Chapter Seven

Summary and Conclusion

The fundamental questions usually asked when discussing the global warming issue are: (1) do we need to reduce CO_2 emissions; (2) and if we do, then by how much; and (3) how will the cuts be partitioned between the developed and developing countries? The only way of answering these questions is by determining the range of possible impacts that may arise from a climate change caused by global warming. Once these impacts have been identified, the next step is to determine if the overall effect on the economic system is positive or negative. For instance, crop yields are expected to increase if atmospheric CO_2 concentrations increase, but yields are also expected to drop from temperature rises caused by the increased CO_2 concentrations. The final result, depending on the overall effect of these two opposing forces will determine whether the impact of a global warming is beneficial or detrimental to the economic system. Therefore, to answer the question posed at the beginning of the chapter, these effects must be made endogenous to the decision making process.

However, most economic studies to date have concentrated on determining the effect a certain percentage reduction in CO_2 emissions will have on economic productivity. The tool which these studies have used in analyzing the impacts have been Computable General Equilibrium (CGE) models (Jorgenson and Wilcoxen 1990, Ghosh 1990). These models are ideal tools for addressing the issues on which market tools should be used to achieve set CO_2 emission levels and the efficiency criteria of these tools. However, these models do not address the issues posed at the beginning of this chapter. Therefore, before we address the issue of market tools to achieve CO_2 reductions, the first step is to identify the level of CO_2 emissions by conducting a study of trade-offs between environmental degradation and economic performance.

This study addresses the questions posed at the beginning of the chapter; i.e., (1) is a reduction in CO_2 emissions warranted; and (2) if yes, by how much must present CO_2 emission levels be reduced; and (3) how will the reduction in CO_2 emissions be partitioned between the two regions. The holistic methodology has the advantage of deriving a solution which compares the costs incurred by the CO_2 emission reduction policies versus damage costs caused by a climate change. If the former is larger than the latter, then it makes little economic sense to embark on CO_2 reduction policies; in other words, the cure is worse than the illness. Furthermore,, to maintain consistency with environmental problems, the time frame of the planning period must be sufficiently long enough to capture the global warming phenomena.

Chapter One gave the necessary background information on the global warming issue and the corresponding feedback effects a climate change can have on the economic system; in particular agriculture yields. This was followed by a discussion in Chapter Two on the policy coordination and cooperation issues which may arise if a global reduction in CO_2 emissions is necessary. Once the basic premises of the problem had been laid down, the problem was restated in mathematical terms.

A detailed description of the multi-region, multi-sectorial, and multi-process model is given in Chapter Three. The formulation style used to endogenize the feedback effects of a climate change on the productivity of the economic system were also discussed in this chapter. The model determines levels of gross domestic product (GDP), consumption, investment, and CO_2 emissions growth rates for each region over a planning period of 50 years. The model results portray the best solution after a complete analysis of trade-offs between costs incurred for transforming an economy from a high to low CO_2 emitting economy and the damage costs incurred if a global warming occurs has been conducted.

Chapter Three also described the process by which the holistic optimal growth model could be modified so as to reflect results which would have resulted if traditional policy analysis was used to address the global warming issue. Traditional policy analysis was defined to be situations in which arbitrary cuts in CO2 emissions are enforced and the corresponding impact of these cuts on economic performance are then analyzed. This means that feedback effects between climate change and the economic system are ignored under traditional analysis conditions.

The data used in the model are presented in Chapter Four. Some of the difficulties faced in calculating the parameters of the model were also discussed in this chapter.

In Chapter Five, five simulation experiments were conducted. The first experiment was conducted on the economic model; the primary purpose was to check the consistency of the model results. This was achieved to a reasonable degree since the rate of increase in atmospheric CO2 concentrations observed in the study was in agreement with results projected by the Inter Governmental Panel on Climate Change (IPCC 1990).

The base experiment also highlighted the importance of land as a factor input for production in the agriculture sector. This result has important consequences for future food supply as it implicitly states that the demand for food by an increasing population cannot be met unless either the productivity of existing land is improved or the earth's supply of forest land is converted for agriculture uses.

Once the validity of the model was established, experiments reflecting present policy options were conducted on the traditional growth model version[1]. The first experiment simulated a situation in which both regions were required to

[1]The traditional version consists of the holistic model but without the feedback equations i.e., no relationship between climate change and the economic system.

restrict CO2 emissions to 1.5 Gt a year. The second experiment reflected a scenario in which only the developed region was required to restrict its CO2 emissions to 1.5 Gt a year. In the third experiment, the developing region was required to restrain its CO2 emissions to 1.5 Gt a year. We called these experiments SCE, PB, and B respectively. The common denominator among these three policy options is that cuts in CO2 emissions were exogenously set by the regions.

The results from the three experiments above were then compared with the results from the holistic model. In the case of the holistic model, the level of CO2 cuts is determined endogenously by the model. Two main conclusions were made from the comparisons. First, the results from the traditional model were inefficient when compared with the holistic model results. Second, the developed region stood to gain if traditional models were used to analyze the global warming issue.

The results from the holistic model indicated that even without an official call for a reduction in CO2 emissions, each region automatically cuts its CO2 emission because the damage incurred by a climate change far outweighs the costs of transforming the economy from a high CO2 to a low CO2 emitting economy. A key result from the Holistic Base is the identification of the level of CO2 emissions by each region in each time period. Rather than having a CO2 emission level set exogenously within the model structure, the methodology used in the holistic model determines the CO2 emission level endogenously by the solution process. Figure 7.1 shows the global level of CO2 emissions observed under Holistic Base conditions.

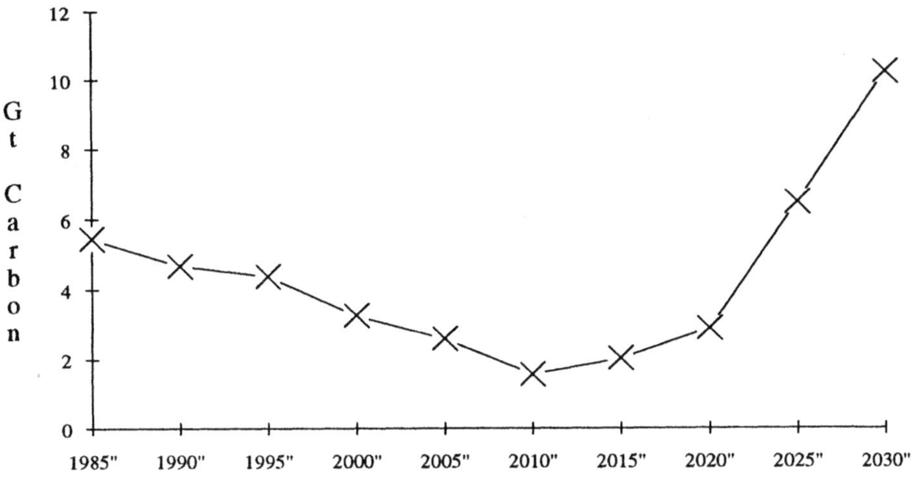

Figure 7.1 Global CO2 emissions under Holistic Base conditions.

Global CO2 emission levels after 2025 should be ignored as the emissions after this time period do not play an integral part in the feedback relationships due to the lag structure used in the feedback equations. Figure 7.2 shows the regional CO2 emissions in the Holistic Base experiment.

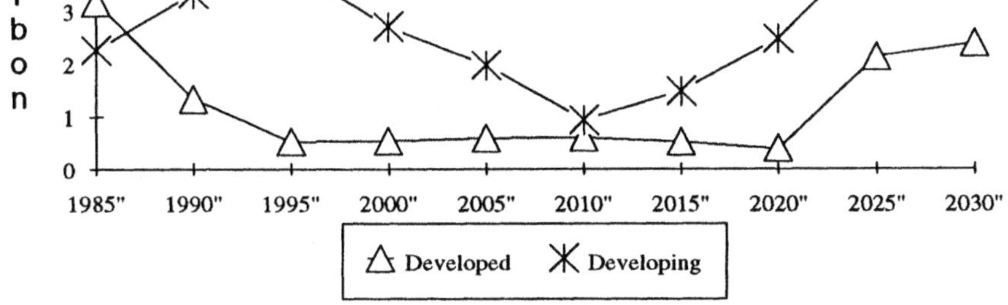

Figure 7.2 Regional CO2 emissions in Holistic Base experiment.

The underlying reason for the developed region to cut its CO2 emissions by a larger amount than that of the developing region can be attributed to the time preference of cutting CO2 emissions to minimize global warming. It is postulated by scientists that the earlier CO2 emissions are cut, the lower will be the degree of global warming. Therefore, as the major emitters of CO2 presently are the developed countries, the most efficient way of curbing the effects of global warming calls for CO2 reductions by the developed region. This then places lower pressure on the developing region to transform its economy to a low CO2 emitting economy. Figure 7.3 below shows the total level of CO2 emissions by the two regions under the two regimes over the planning period.

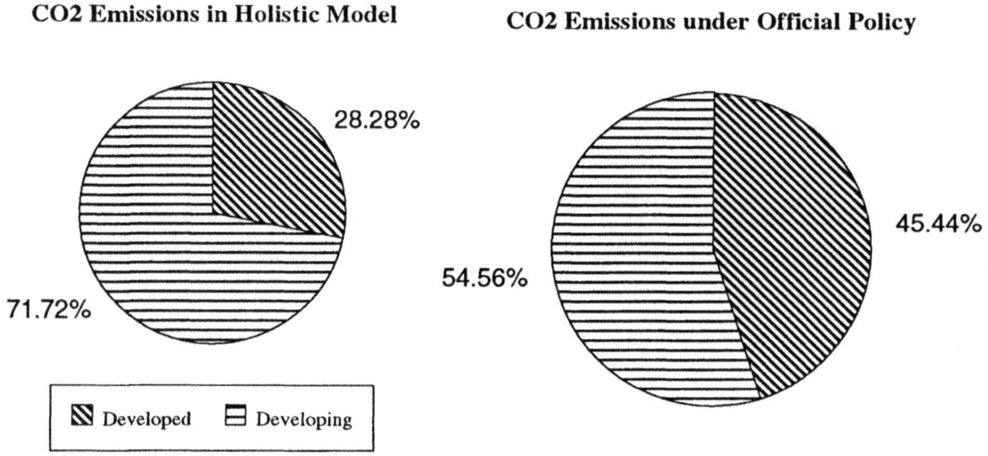

Figure 7.3 CO2 emissions under holistic and official policy conditions[2].

We can infer from the results of the holistic model, that a global CO2 emission of greater than 217 Gt over the next 50 years will cause significant negative feedback effects on the economic system. The 0.64 degree rise observed in the holistic case is the threshold level of temperature rise which gives the best trade-off between damage caused by feedback effects of climate change and the

[2]Official policy refers to the SCE policy.

drop in economic performance caused by the switch from high to low emitting CO2 technology; temperature increases above this value will cause significant drop in economic activity. In the PB scenario, a temperature rise of 1 degree was observed. As this level is larger than the minimum level, the feedback effects will cause significant reductions in capital and land productivity. The base experiment had already highlighted the importance of land and any factor causing a drop in land productivity will cause significant negative impacts on the economic system.

In the B experiment the developing region is being asked to cut its CO2 emissions but given aid for adopting this policy. In this case the results indicate inefficient outcomes as the temperature rise witnessed under this scenario is still again higher than the minimum level of the holistic base. This will again result in a situation similar to the PB scenario. From the results observed in SCE, PB, and B, it can be inferred that the only possibility of these traditional methods setting arbitrary cuts in CO2 emission levels that will achieve a solution superior to that of the holistic model is: (1) when the economic variables fare better than in holistic; and (2) when allowing a temperature rise which is lower than that determined by the holistic model.

As the results from Chapter Five were determined to a large extent by the feedback parameters in the feedback equations, sensitivity analysis was conducted on theses variables. Results from these experiments were presented in Chapter Six. The results indicated that if a temperature rise of two degrees decreases agriculture yields by 10 percent while the doubling of atmospheric carbon dioxide increased yields by 30 and 10 percent in the developed and developing regions respectively, then policymakers do not need to take any action to curb the present trend of CO2 emissions. The feedback effects were not significant enough to warrant any action to reduce global CO2 emissions. Numerous studies have been conducted to investigate the feedback effects of climate change on agriculture yields. However, no conclusive results have been established between these two variables and until there is conclusive evidence of the relationship, economists have to rely on plausible ranges over which the relationship may hold.

As in any modeling experience, there are limitations in the model which can be overcome with further research. Prices are not modeled explicitly in the model structure. The inclusion of prices would have increased the computational complexity considerably. A decision was made in the early stages of model development that although prices play an important role that its contribution in this particular study would be small when compared with the increase in computational complexity it adds to the model solving process. However, rather than completely ignore the role prices play, we decided to capture the effects of the price mechanism within the model structure in another manner without increasing the computational complexity.

The main reason prices are used in a majority of the present generation of global warming models is to capture price induced substitution effect between various fuel types to meet energy demand. In a majority of models, the prices of fuel types over the planning horizon are exogenously specified. Then emission target levels are specified. The models are than solved to obtain carbon tax rates which achieve these target levels. Due to the different carbon content of the different fossil fuels, depending on the emission target level and the forecasted prices, tax rates are determined. The final price, i.e., the forecast price plus the tax rate then determine the degree of substitution between the fuels to satisfy final energy demand in the economy. We decided to capture the substitution process through the use of emission protocols and the capital output coefficients. The former works in the following manner; for example, by placing a 50 percent emission protocol on CO2, the price of coal implicitly goes up with respect to the price of other fuels with a lower carbon content and it is this mechanism which causes substitution between the fuels. The capital-output coefficients on the other hand, capture the difference in the basic price structure between the fuel types. Although the price of the fuel is only part of the coefficient, we believe that by changing the capital-output coefficient over time, we capture the price change of major fuels to an approximation which is sufficient for the purposes of this study.

We do admit that this style of modeling the price mechanism is not as rigorous as a model structure with prices explicitly specified within the model structure as an endogenous variable. However, we should like to stress at this point that the main focus of the study is to identify the optimal level of CO2 emissions which gives the best trade-off position between economic performance and environmental damage. The inclusion of prices may have improved the magnitude resolution of the results but the qualitative results would have remained the same. However, we should like to point out that a majority of the models which use prices rely on forecasted fuel prices over the planning period. Forecasting fuel prices over a period of 50 years would just add another degree of uncertainty to a problem which is plagued by uncertainty.

The model structure is flexible for users to modify and add these price variables in the future. However, if the policy maker wants to capture the effect of dwindling oil and gas supplies, he or she can easily implement this by changing the capital-output coefficients appropriately over the planning horizon.

Another area for potential future research is to include the effect of the other greenhouse gases into the holistic model to determine the most cost efficient method to reduce the feedback effects of climate change. As each greenhouse gas has a different global warming potential, it may be cheaper to reduce methane or cholofluorocarbons which have a higher warming potential than that of CO2. Another avenue for future research is the incorporation of other significant feedback effects which this study ignored; for example, the effect of sea-level rises on coastal zones.

Our goal in this study was to develop a model of global warming which included both the potential conflict between the developing and developed countries and the feedback effects of atmospheric CO_2 concentration and temperature rises on economic productivity. Once the model was developed, it was then used to illustrate the kinds of policy options that must be faced in the future and to focus on the data and model specifications which are crucial to the outcome.

We have found that the following data and specifications are of importance:

1) Land-output coefficient and constraints.

2) Parameters in the capital and land equations which capture the effects of atmospheric CO_2 concentration and temperature rises.

We have also found that the following policy issues are of substantial importance:

1) If the feedback effects of a climate change are taken into account in the policy making process, then the decision to reduce CO_2 emissions as well as the amount of reduction is determined endogenously within the decision process.

2) The equitable division of future CO_2 emissions between the developing and developed countries is crucial to the success of a CO_2 emission reduction policy.

We should like to emphasize that the results presented in this paper are not intended to be policy recommendations, but more of an illustration of the conceptual and stylistic approach to economic-climate change modeling. The structure of the mathematical model highlights the importance of capturing spatial, temporal, and system linkages within a single model structure.

Appendix ONE

The GAMS Statement

The GAMS program which contains the mathematical model presented in Chapter Three is presented below. There are three consecutive solve statements, the first solve statement solves a traditional economic growth model, the second solve statement solves the economic model with the carbon cycle and temperature model included. The third and final solve statement solves the holistic model structure; i.e., the feedback equations are included in the model.

The GAMS Input File

```
*This holistic economic growth model consists of the economic sub-system with
*the aid equations, the carbon cycle model, the global surface temperature model,
*and the feedback equations. This model was developed by Anantha K.
*Duraiappah of the Economics Department,  University of Texas at Austin,
*Texas, 1991.

$OFFSYMXREF OFFSYMLIST OFFUELLIST OFFUELXREF

        SETS
        R   REGIONS   /Deved, Deveing /
*    Deved represents the developed countries
*    Deveing represents the developing countries

        RD(R)  Region which provides aid /Deved /
        RR(R) Region which receives aid  /Deveing /

        S   SECTORS   /Industry, Agri, Services/
*       Agri stands for the agriculture sector
*       Industry represents manufacturing, construction and other heavy
industries
*       Services represents communications, transport, and other light industries
```

SR(S) Energy dependent sectors /Industry, Services /

P PROCESS /Abat,Inter,Intens /
* Abat represents pollution abatement processes
* Inter represents pollution technique that falls in between Abat and Intens
* Intens represents pollution intensive techniques

RI NUMBERS /1*4/ These numbers represent the natural carbon reservoirs

T TIME /1985,1990,1995,2000,2005,2010,2015,2020,2025,2030, 2035 /

TE(T) SUBSET
/1985,1990,1995,2000,2005,2010,2015,2020,2025,2030/

TB(T) BASE YEAR
TT(T) TERMINAL YEAR ;

TB(T) =YES$(ORD(T) EQ 1) ;
TT(T) =YES$(ORD(T) EQ CARD(T)) ;

ALIAS (S,SP) ;
ALIAS (R,RP) ;

TABLE Alphar(R,S,SP) INPUT-OUTPUT MATRIX

	Industry	Agri	Services
Deved.Industry	0.33	0.08	0.11
Deved.Agri	0.05	0.34	0.03
Deved.Services	0.14	0.12	0.20
Deveing.Industry	0.30	0.07	0.07
Deveing.Agri	0.10	0.17	0.01
Deveing.Services	0.18	0.14	0.14

TABLE Delta(R,S) DEPRECIATION RATES

	Industry	Agri	Services
Deved	0.02	0.02	0.02
Deveing	0.02	0.02	0.02

TABLE Beta(S,SP) CAPITAL COEFFICIENTS

	Industry	Agri	Services
Industry	1.39	0.69	0.35
Agri	0.00	0.00	0.00
Services	0.00	0.00	0.00

* The parameter Betar assigns the same values of beta for both regions
PARAMETER Betar(R,S,SP) ;
Betar(R,S,SP) = Beta(S,SP) ;

TABLE Zeta(R,S,*) Energy unit (Exajoules) required to produce unit output in base year

	1985
Deved.Industry	0.016
Deved.Agri	0.0
Deved.Services	0.009
Deveing.Industry	0.034
Deveing.Agri	0.0
Deveing.Services	0.016

SCALAR EFF Efficiency parameter for energy use in sectors /0.99/ ;

SCALAR TAU Years per time period /5/ ;

PARAMETER ZETAT (R,S,T) Energy unit (Exajoules) in each subsequent time period ;

ZETAT(R,S,T) = ZETA(R,S,"1985")*(EFF**(TAU*(ORD(T)-1))) ;
DISPLAY ZETAT ;

TABLE ZETACO2 (S,P) CO2 emission coefficients for each process

	Abat	Inter	Intens
Industry	0.00	0.0177	0.0238
Agri	0.00	0.00	0.00
Services	0.00	0.0177	0.0238

PARAMETER Zetaf(R) CO2 emissions from deforestation
/Deved 0.06
Deveing 0.12 / ;

PARAMETER Theta(RI) Transfer rates of CO2 from natural reservoirs
/1 0.049
2 0.136
3 0.123
4 0.09 / ;

TABLE zsini(R,S,*) Initial values for each region's sectorial GDP levels in 1980 billion dollars

	1985
Deved.Industry	3500
Deved.Agri	1000
Deved.Services	5500
Deveing.Industry	816
Deveing.Agri	840
Deveing.Services	696

TABLE ZSGROWTH(R,S) Sectorial growth rate in each region

	Industry	Agri	Services
Deved	1.03	1.03	1.03
Deveing	1.07	1.04	1.02

PARAMETER zstilde(R,S,T) Desired path for sectorial GDP ;

zstilde(R,S,T) = zsini(R,S,"1985") *(ZSGROWTH(R,S)**
(TAU*(ORD(T)-1)));

TABLE qsini(R,S,*) Initial values for sectorial output in base year in 1980 billion dollars

	1985
Deved.Industry	6900
Deved.Agri	2420
Deved.Services	8440
Deveing.Industry	1415
Deveing.Agri	1200
Deveing.Services	1300

TABLE QSGROWTH(R,S) Growth rate for sectorial output

	Industry	Agri	Services
Deved	1.03	1.03	1.03
Deveing	1.07	1.04	1.02

PARAMETER QSTILDE(R,SP,T) Desired path for sectorial output;
qstilde(R,S,T) = qsini(R,S,"1985") *(QSGROWTH(R,S)**
(TAU*(ORD(T)-1)));

TABLE Wzs(R,S,T) Penalty matrix for sectorial GDP

	1985	1990	1995	2000	2005	2010	2015	2020
Deved.Industry	30	30	30	30	30	30	30	30
Deved.Agri	300	300	300	300	300	300	300	300
Deved.Services	50	50	50	50	50	50	50	50
Deveing.Industry	460	460	460	460	460	460	460	460
Deveing.Agri	430	430	430	430	430	430	430	430
Deveing.Services	620	620	620	620	620	620	620	620

+	2025	2030	2035
Deved.Industry	30	30	30
Deved.Agri	300	300	300
Deved.Services	50	50	50
Deveing.Industry	460	460	460
Deveing.Agri	430	430	430
Deveing.Services	620	620	620

TABLE MPC(R,S) Average propensity to consume

	Industry	Agri	Services
Deved	0.40	0.35	0.6
Deveing	0.30	0.50	0.50

TABLE CGROWTH(R,*) Growth rate for the agriculture sector

	Agri
Deved	1.005
Deveing	1.04

```
PARAMETER ctilde(R,S,T) Desired path for sectorial consumption ;
ctilde(R,SR,T) = MPC(R,SR) * qstilde(R,SR,T) ;

ctilde(R,"Agri","1985") = MPC(R,"Agri")*qstilde(R,"Agri","1985") ;

ctilde(R,"Agri",T) = ctilde(R,"Agri","1985")*
(CGROWTH(R,"Agri")**(TAU*(ORD(T)-1))) ;
```

TABLE Wc(R,S,T) Priority matrix for consumption variables

	1985	1990	1995	2000	2005	2010	2015	2020
Deved.Industry	40	40	40	40	40	40	40	40
Deved.Agri	121	121	121	121	121	121	121	121
Deved.Services	4	4	4	4	4	4	4	4
Deveing.Industry	2000	2000	2000	2000	2000	2000	2000	2000
Deveing.Agri	170	170	170	170	170	170	170	170
Deveing.Services	250	250	250	250	250	250	250	250

+	2025	2030	2035
Deved.Industry	40	40	40
Deved.Agri	121	121	121
Deved.Services	4	4	4
Deveing.Industry	2000	2000	2000
Deveing.Agri	170	170	170
Deveing.Services	250	250	250

PARAMETER BO(R) Base level of forest land in million hectares

/Deved 1824
Deveing 2265 / ;

TABLE DO(*,R) Initial conditions for agriculture land in million hectares

	Deved	Deveing
1985	670	800

TABLE IO(R,S,*) Initial values for sectorial investment in 1980 billion dollars

	1985
Deved.Industry	500
Deved.Agri	300
Deved.Services	500
Deveing.Industry	340
Deveing.Agri	150
Deveing.Services	250

PARAMETER dtilde(R,T) Desired path for agriculture land usage ;

PARAMETER LU(R) Land usage growth rate
/Deved 1.0032
Deveing 1.012 / ;

dtilde(R,T) = DO("1985",R)*(LU(R) **(TAU*(ORD(T)-1))) ;

TABLE Wd(R,T) Priority matrix for agriculture land usage

	1985	1990	1995	2000	2005	2010
Deved	3E+4	3E+4	3E+4	3E+4	3E+4	3E+4
Deveing	2E+5	2E+5	2E+5	2E+5	2E+5	2E+5

+	2015	2020	2025	2030	2035
Deved	3E+4	3E+4	3E+4	3E+4	3E+4
Deveing	2E+5	2E+5	2E+5	2E+5	2E+5

PARAMETER natilde(R,T) Desired path for regional CO2 emissions in carbon Gt per year ;
natilde("Deved,T) = 1.5 ;
natilde("Deveing",T) = 1.5 ;

TABLE Wna(R,T) Penalty matrix for regional CO2 emission variable

	1985	1990	1995	2000	2005	2010
Deved	7E-8	7E-8	7E-8	7E-8	7E-8	7E-8
Deveing	7E-8	7E-8	7E-8	7E-8	7E-8	7E-8

+	2015	2020	2025	2030	2035
Deved	7E-8	7E-8	7E-8	7E-8	7E-8
Deveing	7E-8	7E-8	7E-8	7E-8	7E-8

TABLE IGROWTH(R,S) growth rates for sectorial investment

	Industry	Agri	Services
Deved	1.02	1.03	1.03
Deveing	1.04	1.03	1.02

```
PARAMETER itilde(R,S,T) DESIRED PATH FOR INVESTMENT ;
itilde(R,S,T) = io(R,S,"1985") *(iGROWTH(R,S)
**(TAU*(ORD(T)-1)));
```

TABLE Lambdai(R,S,T) Penalty matrix for investment variables

	1985	1990	1995	2000	2005	2010	2015	2020
Deved.Industry	120	120	120	120	120	120	120	120
Deved.Agri	340	340	340	340	340	340	340	340
Deved.Services	120	120	120	120	120	120	120	120
Deveing.Industry	260	260	260	260	260	260	260	260
Deveing.Agri	1300	1300	1300	1300	1300	1300	1300	1300
Deveing.Services	500	500	500	500	500	500	500	500

+	2025	2030	2035
Deved.Industry	120	120	120
Deved.Agri	340	340	340
Deved.Services	120	120	120
Deveing.Industry	260	260	260
Deveing.Agri	1300	1300	1300
Deveing.Services	500	500	500

```
PARAMETER ftilde(R,T) Desired path for deforestation ;

PARAMETER DF(R) Desired rate of deforestation per year

        /Deved      1
         Deveing    11  / ;

         ftilde(R,T) =DF(R) ;
```

TABLE Lambdaf(R,T) Penalty matrix for deforestation variable

	1985	1990	1995	2000	2005	2010
Deved	3.5E+3	3.5E+3	3.5E+3	3.5E+3	3.5E+3	3.5E+3
Deveing	3.0E+8	3.0E+8	3.0E+3	3.0E+3	3.0E+3	3.0E+3

+	2015	2020	2025	2030	2035
Deved	3.5E+3	3.5E+3	3.5E+3	3.5E+3	3.5E+3
Deveing	3.0E+3	3.0E+3	3.0E+3	3.0E+3	3.0E+3

TABLE MUK(R,S,P) Constant term in capital-output variable

	Abat	Inter	Intens
Deved.Industry	3.0	1.68	1.40
Deved.Agri	1.08	1.20	1.32
Deved.Services	2.50	1.50	1.25
Deveing.Industry	4.87	2.74	2.11
Deveing.Agri	1.20	1.33	1.46
Deveing.Services	3.59	2.15	1.65

TABLE TECK(R,S,P) Technical innovation factor for constant term in capital-output variable

	Abat	Inter	Intens
Deved.Industry	0.99	1.00	1.00
Deved.Agri	1.00	1.00	1.00
Deved.Services	0.99	1.00	1.00
Deveing.Industry	0.99	1.00	1.00
Deveing.Agri	1.00	1.00	1.00
Deveing.Services	0.99	1.00	1.00

PARAMETER MUKT(R,S,P,T) Capital-output coefficients reflecting technical innovation ;
MUKT(R,S,P,T) = MUK(R,S,P)* (TECK(R,S,P)**(TAU*(ORD(T)-1)));
DISPLAY MUKT ;

TABLE NUK(R,S,P) Temperature feedback effect on capital-output coefficient

	Abat	Inter	Intens
Deved.Industry	0	0	0
Deved.Agri	0.36	0.40	0.44
Deved.Services	0	0	0
Deveing.Industry	0	0	0

Deveing.Agri	0.40	0.44	0.48
Deveing.Services	0	0	0

TABLE SIGMAK(R,S,P) CO2 fertilization effect on capital-output variable

	Abat	Inter	Intens
Deved.Industry	0	0	0
Deved.Agri	-0.0009	-0.001	-0.001
Deved.Services	0	0	0
Deveing.Industry	0	0	0
Deveing.Agri	-0.0004	-0.0004	-0.0005
Deveing.Services	0	0	0

TABLE MUD(R,S,P) Constant term in land-output variable

	Abat	Inter	Intens
Deved.Industry	0	0	0
Deved.Agri	1.00	0.67	0.28
Deved.Services	0	0	0
Deveing.Industry	0	0	0
Deveing.Agri	0.95	0.67	0.28
Deveing.Services	0	0	0

TABLE TECD(R,S,P) Technical innovation factor for constant term in land-output variable

	Abat	Inter	Intens
Deved.Industry	1.00	1.00	1.00
Deved.Agri	0.99	0.99	0.99
Deved.Services	1.00	1.00	1.00
Deveing.Industry	1.00	1.00	1.00
Deveing.Agri	0.99	0.99	0.99
Deveing.Services	1.00	1.00	1.00

PARAMETER MUDT(R,S,P,T) Land-output coefficients reflecting technical innovation ;
MUDT(R,S,P,T) = MUD(R,S,P)* (TECD(R,S,P)**(TAU*(ORD(T)-1)));
DISPLAY MUDT ;

TABLE NUD(R,S,P) Temperature feedback effect on land-output coefficient

	Abat	Inter	Intens
Deved.Industry	0	0	0
Deved.Agri	0.33	0.22	0.09
Deved.Services	0	0	0
Deveing.Industry	0	0	0
Deveing.Agri	0.32	0.22	0.09
Deveing.Services	0	0	0

TABLE SIGMAD(R,S,P) CO2 fertilization effect on land-output variable

	Abat	Inter	Intens
Deved.Industry	0	0	0
Deved.Agri	-0.0008	-0.0005	-0.0002
Deved.Services	0	0	0
Deveing.Industry	0	0	0
Deveing.Agri	-0.0003	-0.0002	-0.0001
Deveing.Services	0	0	0

VARIABLES

k(t,r,s)	Capital Stocks
d(t,r)	Land Supply
i(t,r,s)	Investment levels
c(T,R,S)	Consumption levels
hr(t,rr,s)	Sectorial foreign aid received by recipient region
hd(t,rd,s)	Sectorial foreign aid sent by donor region
faid(t,r)	Regional foreign aid
f(t,r)	Deforestation rate
qp(t,r,s,p)	Output at process level
qs(t,r,s)	Output at sectorial level
zs(t,r,s)	Sectorial gross domestic product
zr(t,r)	Regional gross domestic product

qj(t,r,s,p)	Energy units used at process level
e(t,r,s,p)	CO2 emissions in Gt at the process level
er(t,r)	CO2 emissions in Gt by regions
eg(t)	Global CO2 emissions in Gt
J	Criterion

POSITIVE VARIABLES
i,ip,c,k,d,f,qp,qs,qj,e,er,eg,zs,zr,hd,hr ;

EQUATIONS

CRITERION	Objective function
MATBALD(t,rd,s)	Material balance for donor region
MATBALR(t,rr,s)	Material balance for recipient region
CAPITAL(t,r,s)	Capital accumulation
SECTORIAL(t,r,s)	Output at sector level
GDPS(t,r,s)	Sectorial GDP
GDP(t,r)	Regional GDP
CAPSTOCK(t,r,s)	Capital stock constraint
BTUC(t,r,s,p)	Energy demand by processes
POLLUTION(t,r,s,p)	CO2 emission by processes (fossil fuels)
RPOLL(t,r)	CO2 emission by regions (fossil fuels and deforestation)
GLOBPOLL(t)	Global CO2 emissions
TAID(t)	Total aid balance
UAID(t)	Upper bound on aid from donor region
AIDC(t,rr,s)	Foreign aid condition
UPPERF(t,r)	Deforestation constraint in each time period
LOWERP(t,r,s,p)	Lower bound on output by pollution intermediate process
LAND(t,r)	Land constraint
LANDSUPP(t,r)	Land supply
DEFOREST(r)	Deforestation constraint over planning period ;

```
*       OBJECTIVE FUNCTION

CRITERION..  J    =E=    0.5*SUM((t,r,s),
                         (zs(t,r,s)-zstilde(r,s,t))
                         *Wzs(r,s,t)
                         *(zs(t,r,s)-zstilde(r,s,t)))

                         + 0.5* SUM((t,r,s),
                         (c(t,r,s)-ctilde(r,s,t))
                         *Wc(r,s,t)
                         *(c(t,r,s)-ctilde(r,s,t)))

                         + 0.5*SUM((t,r),
                         (d(t,r)-dtilde(r,t))
                         *wd(r,t)
                         *(d(t,r)-dtilde(r,t)))

                         + 0.5*SUM((t,r,s),
                         (i(t,r,s)-itilde(r,s,t))
                         *Lambdai(r,s,t)
                         *(i(t,r,s)-itilde(r,s,t))) ;


*       MATERIAL BALANCE FOR THE DONOR REGION

 MATBALD(t,rd,s)..  qs(t,rd,s)  =E= SUM(sp,Alphar(rd,s,sp)*
 qs(t,rd,sp)) + SUM(sp,Betar(rd,s,sp)*i(t,rd,sp)) + c(t,rd,s)+
 hd(t,rd,s) ;

*       MATERIAL BALANCE FOR THE RECIPIENT REGION

 MATBALR(t,rr,s)..  qs(t,rr,s)  =E= SUM(sp,Alphar(rr,s,sp)*
 qs(t,rr,sp)) + SUM(sp,Betar(rr,s,sp)*i(t,rr,sp)) + c(t,rr,s)
 -hr(t,rr,s) ;

*       CAPITAL ACCUMULATION

 CAPITAL(t+1,r,s)..  k(t+1,r,s) =E=(1-DELTA(r,s))** TAU
 * k(t,r,s)   + TAU * i(t,r,s) ;
```

```
*       OUTPUT AT SECTOR LEVEL

SECTORIAL(t,r,s).. qs(t,r,s) =E= SUM(p,qp(t,r,s,p)) ;

*       SECTORIAL GDP

GDPS(t,r,s)..  zs(t,r,s) =E=  qs(t,r,s) - SUM(sp, ALPHAR(r,s,sp)*qs(t,r,sp));

*       REGIONAL GDP

GDP(t,r)..   zr(t,r) =E= SUM(s,zs(t,r,s)) ;

*       CAPITAL STOCK CONSTRAINT

CAPSTOCK(t,r,s).. SUM(p, MUKT(r,s,p,t)*qp(t,r,s,p)) =L=  k(t,r,s) ;

*       ENERGY DEMAND BY PROCESSES

BTUC(t,r,s,p)..  qj(t,r,s,p) =E= qp(t,r,s,p)*ZETAT(r,s,t) ;

*       CO2 EMISSION BY PROCESSES (FOSSIL FUELS)

POLLUTION(t,r,s,p)..  e(t,r,s,p) =E= (qj(t,r,s,p) * ZETACO2(s,p));

*       CO2 EMISSIONS AT REGIONAL LEVEL (DEFORESTATION
*       AND FOSSIL FUELS)

RPOLL(t,r).. er(t,r) =E= SUM((s,p),e(t,r,s,p)) + ZETAF(r)*f(t,r) ;

*       GLOBAL CO2 EMISSIONS

GLOBPOLL(t).. eg(t) =E= SUM(r,er(t,r)) ;

*       TOTAL AID BALANCE

TAID(t)..  SUM((rd,s),HD(t,rd,s)) =E= SUM((rr,sr),HR(t,rr,sr)) ;
```

```
*       UPPER BOUND ON AID FROM DONOR REGION

 UAID(t).. SUM((rd,s),HD(t,rd,s)) =L= 0.15* SUM(rd,zr(t,rd)) ;

*       FOREIGN AID CONDITION

AIDC(t,rr,sr).. hr(t,rr,sr) =L= (MUKT(rr,rs,"Abat",t)-
MUKT(rr,sr,"Intens",t))*qp(t,rr,sr,"Abat") +
(MUKT(rr,sr,"Inter",t)- MUKT(rr,sr,"Intens",t))*qp(t,rr,sr,"Inter");


*       DEFORESTATION CONSTRAINT IN EACH TIME PERIOD

UPPERF(t,r).. f(t,r) =L= 20 ;

*       LOWER BOUND ON OUTPUT BY POLLUTION
*       INTERMEDIATE PROCESS

LOWERP(t,r,sr,"Inter").. qp(t,r,sr,"Inter") =G= 0.5*qs(t,r,sr);


*       LAND CONSTRAINT

LAND(t,r).. SUM((s,p), MUD(r,s,p)*qp(t,r,s,p)) =L= d(t,r) ;

*       LAND SUPPLY

LANDSUPP(t+1,r).. d(t+1,r) =E= d(t,r) + TAU*f(t,r) ;

*       DEFORESTATION CONSTRAINT OVER PLANNING PERIOD

DEFOREST(r).. SUM(t, TAU*f(t,r)) =L= BO(r) ;



* The following option statements were used to solve the model
OPTION ITERLIM = 50000;
OPTION NLP=MINOS5 ;
OPTION LIMCOL = 0 ;
```

```
OPTION LIMROW = 0    ;
OPTION SOLPRINT = Off ;
OPTION DECIMALS = 5 ;
OPTION RESLIM = 10000 ;

*        These are fixed initial values for capital stock in base year

 k.FX ("1985","Deved","Energy") = 10916;
 k.FX ("1985","Deved","Agri") = 3195 ;
 k.FX ("1985","Deved","Services") = 11922   ;
 k.FX ("1985","Deveing","Energy") = 3458 ;
 k.FX ("1985","Deveing","Agri") =1440;
 k.FX ("1985","Deveing","Services") =2489   ;

*        These are initial values for sectorial GDP in base year

  zs.L("1985","Deved","Energy") = 3500 ;
  zs.L("1985","Deved","Agri") = 1000 ;
  zs.L("1985","Deved","Services") = 5500 ;
  zs.L("1985","Deveing","Energy") = 816 ;
  zs.L("1985","Deveing","Agri") = 840 ;
  zs.L("1985","Deveing","Services") = 696 ;

*        This is fixed initial value for agricultural land in base year

  d.FX  ("1985",R) = DO("1985",R) ;


*        Foreign aid is only used by the industry and service sectors in the
*        recipient region

  hr.FX(T,rr,"Agri") = 0.0 ;

* The results from the first solve statement are used as reference points to help
* find feasible solutions for the non-linear Holistic Base model.

 MODEL ENVIRO /ALL / ;

 SOLVE ENVIRO MINIMIZING J USING NLP ;
```

```
*       The following parameters represent the desired paths for the state and
*       control variables in the criterion function

        PARAMETER zstildep(t,r,s)  ;
        zstildep(t,r,s) = zstilde(r,s,t) ;
        OPTION zstildep:3:1:2;  DISPLAY zstildep ;

        PARAMETER ctildep(t,r,s) ;
        ctildep(t,r,s) = ctilde(r,s,t) ;
        OPTION ctildep:3:1:2;   DISPLAY ctildep;


        PARAMETER dtildep(t,r);
        dtildep(t,r) = dtilde(r,t)
        OPTION dtildep:5:1:1;   DISPLAY dtildep;

        PARAMETER itildep(t,r,s);
        itildep(t,r,s) = itilde(r,s,t);
        OPTION itildep:5:1:2; DISPLAY itildep;

        PARAMETER ftildep(t,r) ;
        ftildep(t,r) = ftilde(r,t) ;
        OPTION ftildep:5:1:1;   DISPLAY ftildep;

        PARAMETER qstildep(t,r,sp);
        qstildep(t,r,sp) = qstilde(r,sp,t);
        OPTION qstildep:3:1:2;  DISPLAY qstildep ;


*       THIS IS THE SECOND MODEL WITH THE CARBON CYCLE  AND
*       TEMPERATURE MODEL ADDED TO THE ECONOMIC MODEL


        VARIABLES

        nat(t)          Carbon mass in Gt in atmosphere
        nbt(t)          Carbon mass in Gt in land
        nmt(t)          Carbon mass in Gt in surface ocean
        erpmv(t)        Concentration of CO2 in atmosphere in ppmv
        tg(t)           Global surface temperature
        J1              Criterion
```

POSITIVE VARIABLES
nat,mbt,nmt,tg,erpmv ;

EQUATIONS

CRITERION1	Objective function
AB(t)	Carbon flux between land and the other reservoirs
ABSO(t)	Carbon flux between atmosphere and the other reservoirs
ASODO(t)	Carbon flux between surface ocean and the other reservoirs
CONE (t)	Conversion formula from mass to concentration in atmosphere
TEMP(t)	Global surface temperature change ;

OBJECTIVE

```
CRITERION1.. J1  =E=    0.5*SUM((t,r,s),
                        (zs(t,r,s)-zstilde(r,s,t))
                        *Wzs(r,s,t)
                        *(zs(t,r,s)-zstilde(r,s,t)))

                        + 0.5* SUM((t,r,s),
                        (c(t,r,s)-ctilde(r,s,t))
                        *Wc(r,s,t)
                        *(c(t,r,s)-ctilde(r,s,t)))

                        + 0.5*SUM((t,r),
                        (d(t,r)-dtilde(r,t))
                        *wd(r,t)
                        *(d(t,r)-dtilde(r,t)))

                        + 0.5*SUM((t,r,s),
                        (i(t,r,s)-itilde(r,s,t))
                        *Lambdai(r,s,t)
                        *(i(t,r,s)-itilde(r,s,t)))
```

```
+0.5*SUM((t,r),
(er(t,r)-natilde(r,t))
*Wna(r,t)
*(er(t,r)-natilde(r,t)));
```

```
*       CARBON FLUX BETWEEN LAND AND THE OTHER
*       RESERVOIRS

AB(T+1)..  nbt(T+1) =E=    Theta("2") * nat(T) +(1-theta("1"))* nbt(T) ;

*       CARBON FLUX BETWEEN ATMOSPHERE AND THE OTHER
*       RESERVOIRS

ABSO(T+1)..  nat(T+1) =E=TAU* eg(T) + (1-Theta("2")-Theta("3"))*nat(T)
+Theta("1")*nbt(T)+Theta("4")*nmt(T) ;

*       CARBON FLUX BETWEEN SURFACE OCEAN AND OTHER
*       RESERVOIRS

ASODO(T+1).. nmt(T+1) =E= Theta("3")*nat(T)+(1-Theta("4"))* nmt(T) ;

*       CONVERSION FORMULA FROM MASS TO CONCENTRATION
*       FOR CO2 IN ATMOSPHERE

CONE (T)..  erpmv (T) =E= nat(T) /2.12;

*       GLOBAL SURFACE TEMPERATURE CHANGE

TEMP(T)..   tg(T) =E= 3 *(log(erpmv (t))-log(270))/0.6931

*       The initial values for the mass in each reservoir are fixed in Gt carbon

        nat.fx("1985")  = 760 ;
        nbt.fx("1985")  = 2060;
        nmt.fx("1985") = 1005;
```

```
*       Lower bound is assigned to atmospheric CO2 concentration to avoid non-
*       linear infeasibility problems

 erpmv.LO(t) = 0.001 ;

        OPTION BRATIO= 1 ;
        OPTION SOLPRINT = OFF;
        OPTION LIMROW = 0;

MODEL BASE
/CRITERION1,MATBALD,CAPITAL,SECTORAL,CAPSTOCK,POLLUTION,
GLOBPOLL,LAND,LANDSUPP,DEFOREST,AB,RPOLL,
ABSO,ASODO,TEMP,UPPERF,MATBALR,GDPS,GDP,LOWERP,
BTUC,CONE,TAID,UAID,AIDC  / ;

SOLVE BASE MINIMIZING J1 USING NLP ;


*       The following parameters represent the optimal values for the variables in
*       the model. The 0.00001 is added to ensure that variables which have
*       a value of zero are displayed in the output file. GAMS display option does
*       not show variable which have a zero value.

        PARAMETER kp1(t,r,s) ;
        kp1(t,r,s) = k.L(t,r,s) + 0.00001  ;
        OPTION kp1:5:1:2; DISPLAY kp1 ;

        PARAMETER dp1(t,r) ;
        dp1(t,r) = d.L(t,r) + 0.00001 ;
        OPTION dp1:5:1:1;  DISPLAY dp1 ;

        PARAMETER ip1(t,r,s) ;
        ip1(t,r,s) = i.L(t,r,s) + 0.00001 ;
        OPTION ip1:5:1:2;  DISPLAY ip1 ;

        PARAMETER hrp1(t,rr,s) ;
         hrp1(t,rr,s) = hr.L(t,rr,s) + 0.00001 ;
         OPTION hrp1:5:1:2;  DISPLAY hrp1 ;
```

```
PARAMETER hdp1(t,rd,s) ;
hdp1(t,rd,s) = hd.L(t,rd,s) + 0.00001 ;
OPTION hdp1:5:1:2;  DISPLAY hdp1 ;

PARAMETER fp1(t,r) ;
fp1(t,r) = f.L(t,r) + 0.00001 ;
OPTION fp1:5:1:1;    DISPLAY fp1;

PARAMETER qpp1(t,r,s,p) ;
qpp1(t,r,s,p) = qp.L(t,r,s,p) + 0.00001 ;
OPTION qpp1:5:1:3;  DISPLAY qpp1 ;

PARAMETER qsp1(t,r,s) ;
qsp1(t,r,s) = qs.L(t,r,s) + 0.00001 ;
OPTION qsp1:5:1:2;   DISPLAY qsp1 ;


PARAMETER zsp1(t,r,s) ;
zsp1(t,r,s) = zs.L(t,r,s) + 0.00001 ;
 OPTION zsp1:5:1:2;   DISPLAY zsp1 ;

PARAMETER cp1(t,r,s) ;
cp1(t,r,s) = c.L(t,r,s) + 0.00001 ;
OPTION cp1:5:1:2;   DISPLAY cp1 ;

PARAMETER ep1(t,r,s,p) ;
ep1(t,r,s,p) = e.L(t,r,s,p) + 0.00001 ;
OPTION ep1:5:1:3;   DISPLAY ep1 ;

PARAMETER erp1(t,r) ;
erp1(t,r) =er.L(t,r) + 0.00001 ;
OPTION erp1:5:1:1;    DISPLAY erp1;

PARAMETER egp1(T) ;
egp1(T) = eg.L(T) + 0.00001 ;
DISPLAY egp1 ;

PARAMETER erpmvp1(T) ;
erpmvp1(T) = erpmv.L(T) + 0.00001 ;
 DISPLAY erpmvp1;
```

```
PARAMETER nap1(T) ;
nap1(T) = nat.L(T) + 0.00001 ;
OPTION nap1:5:0:1;  DISPLAY nap1 ;

PARAMETER tgp1(T) ;
 tgp1(T) = tg.L(T) + 0.00001 ;
OPTION tgp1:5:0:1; DISPLAY tgp1 ;

PARAMETER zrp1(T,R) ;
zrp1(T,R) = zr.L(T,R) + 0.00001 ;
OPTION zrp1:5:1:1;  DISPLAY zrp1 ;


*    THE FEEDBACK EFFECTS ARE ADDED IN AT THIS POINT. THIS
*    FINAL VERSION CONSTITUTES THE HOLISTIC MODEL


VARIABLE J2 ;

POSITIVE VARIABLES
kappa, phi  ;


EQUATIONS
CRITERION2                 Objective function
INIC(*,r,s,p)              Initial condition for capital-output variable
INID(*,r,*,p)              Initial condition for land-output variable
LANDDEF(t,r,*,p)           Land-output variable definition
CAPDEF(t,r,s,p)            Capital-output variable definition
CAPSTOCK1(t,r,s)           Capital stock constraint
LAND2(t,r)                 Land constraint ;


*    OBJECTIVE

CRITERION2..  J2  =E=    0.5*SUM((t,r,s),
                         (zs(t,r,s)-zstilde(r,s,t))
                         *Wzs(r,s,t)
                         *(zs(t,r,s)-zstilde(r,s,t)))
```

```
                    + 0.5* SUM((t,r,s),
                    (c(t,r,s)-ctilde(r,s,t))
                    *Wc(r,s,t)
                    *(c(t,r,s)-ctilde(r,s,t)))

                    + 0.5*SUM((t,r),
                    (d(t,r)-dtilde(r,t))
                    *wd(r,t)
                    *(d(t,r)-dtilde(r,t)))

                    + 0.5*SUM((t,r,s),
                    (i(t,r,s)-itilde(r,s,t))
                    *Lambdai(r,s,t)
                    *(i(t,r,s)-itilde(r,s,t)))

                    +0.5*SUM((t,r),
                    (er(t,r)-natilde(r,t))
                    *Wna(r,t)
                    *(er(t,r)-natilde(r,t)));
```

* **LAND-OUPUT VARIABLE DEFINITION**

```
LANDDEF(t+1,r,"Agri",p)..  phi(t+1,r,"Agri",p) =E= MUD(R,"Agri",P)+
NUD(R,"Agri",P)* (tg(T+1)-tg("1985")) +
SIGMAD(R,"Agri",P)*(erpmv(T+1)-erpmv("1985")) ;
```

* **CAPITAL OUTPUT VARIABLE DEFINITION**

```
CAPDEF(t+1,r,s,p)..  kappa(t+1,r,s,p) =E= MUKT(R,S,P,T) +
NUK(R,S,P)* (tg(T+1)-tg("1985")) +
SIGMAK(R,S,P)*(erpmv(T)-erpmv("1985")) ;
```

* **INITIAL CONDITION FOR CAPITAL-OUTPUT VARIABLE**

```
INIC("1990",R,S,P)..  kappa("1985",r,s,p) =E= MUK(R,S,P) ;
```

* **INITIAL CONDITION FOR LAND-OUTPUT VARIABLE**

```
INID("1990",R,"Agri",P)..  phi("1985",R,"Agri",P) =E=  MUD(R,"Agri",P) ;
```

```
*       CAPITAL STOCK CONSTRAINT

 CAPSTOCK1(t,r,s)..  sum(p, kappa(t,r,s,p)*qp(t,r,s,p)) =L=  k(t,r,s) ;

*       LAND CONSTRAINT

  LAND2(t,r)..       SUM(P, phi(t,r,"Agri",p)*qp(t,r,"Agri",p)) =L=  d(t,r) ;


   OPTION BRATIO= 1 ;
   OPTION SOLPRINT = OFF;
   OPTION LIMROW = 0 ;

 MODEL HOLISTIC
/CRITERION2,MATBALD,CAPITAL,SECTORAL,CAPSTOCK1,
 POLLUTION,GLOBPOLL,LAND2,LANDSUPP,DEFOREST,AB,RPOLL,
 ABSO,ASODO,TEMP,UPPERF,MATBALR,CAPDEF,LANDDEF,INID,
 BTUC,CONE,INIC,GDPS,GDP,TAID,UAID,AIDC / ;

  SOLVE HOLISTIC MINIMIZING J2 USING NLP ;


        PARAMETER kp2(t,r,s) ;
        kp2(t,r,s) = k.L(t,r,s) + 0.00001  ;
        OPTION kp2:5:1:2; DISPLAY kp2 ;


        PARAMETER dp2(t,r) ;
        dp2(t,r) = d.L(t,r) + 0.00001 ;
        OPTION dp2:5:1:1;  DISPLAY dp2 ;

        PARAMETER ip2(t,r,s) ;
        ip2(t,r,s) = i.L(t,r,s) + 0.00001 ;
        OPTION ip2:5:1:2;  DISPLAY ip2 ;

        PARAMETER hrp2(t,rr,s) ;
        hrp2(t,rr,s) = hr.L(t,rr,s) + 0.00001 ;
        OPTION hrp2:5:1:2;  DISPLAY hrp2 ;
```

```
PARAMETER hdp2(t,rd,s) ;
hdp2(t,rd,s) = hd.L(t,rd,s) + 0.00001 ;
OPTION hdp2:5:1:2;  DISPLAY hdp2 ;

PARAMETER fp2(t,r) ;
fp2(t,r) = f.L(t,r) + 0.00001 ;
OPTION fp2:5:1:1;   DISPLAY fp2;

PARAMETER qpp2(t,r,s,p) ;
qpp2(t,r,s,p) = qp.L(t,r,s,p) + 0.00001 ;
OPTION qpp2:5:1:3;  DISPLAY qpp2 ;

PARAMETER qsp2(t,r,s) ;
qsp2(t,r,s) = qs.L(t,r,s) + 0.00001 ;
OPTION qsp2:5:1:2;   DISPLAY qsp2 ;

PARAMETER zsp2(t,r,s) ;
zsp2(t,r,s) = zs.L(t,r,s) + 0.00001 ;
OPTION zsp2:5:1:2;   DISPLAY zsp2 ;


PARAMETER cp2(t,r,s) ;
cp2(t,r,s) = c.L(t,r,s) + 0.00001 ;
OPTION cp2:5:1:2;   DISPLAY cp2 ;

PARAMETER ep2(t,r,s,p) ;
ep2(t,r,s,p) = e.L(t,r,s,p) + 0.00001 ;
 OPTION ep2:5:1:3;   DISPLAY ep2 ;

PARAMETER erp2(t,r) ;
erp2(t,r) =er.L(t,r) + 0.00001 ;
OPTION erp2:5:1:1;    DISPLAY erp2;

PARAMETER egp2(T) ;
egp2(T) = eg.L(T) + 0.00001 ;
DISPLAY egp2 ;

PARAMETER erpmvp2(T) ;
erpmvp2(T) = erpmv.L(T) + 0.00001 ;
DISPLAY erpmvp2;
```

```
PARAMETER nap2(T) ;
 nap2(T) = nat.L(T) + 0.00001 ;
OPTION nap2:5:0:1;  DISPLAY nap2 ;

PARAMETER tgp2(T) ;
tgp2(T) = tg.L(T) + 0.00001 ;
OPTION tgp2:5:0:1; DISPLAY tgp2 ;

PARAMETER zrp2(t,r) ;
zrp2(t,r) = zr.L(t,r) + 0.00001 ;
OPTION zrp2:5:1:1;  DISPLAY zrp2 ;

PARAMETER Kappap(t,r,s,p) ;
Kappap(t,r,s,p) = Kappa.L(t,r,s,p) + 0.00001;
OPTION Kappap:5:3:1; DISPLAY Kappap ;

PARAMETER Phip(t,r,s,p) ;
Phip(t,r,s,p) = Phi.L(t,r,s,p) + 0.00001;
OPTION Phip:5:3:1; DISPLAY Phip;
```

Appendix TWO

Data Calibration

This appendix illustrates the process by which the constant term in the capital-output and land-output equations for the processes in the three sectors in the respective regions were caliberated.

Constant term in capital-output variable

Capital-output coefficients for countries are difficult to obtain. It becomes a more difficult task when the economy is aggregated to three sectors and even further when the world is divided into two regions. Rather than embarking on the task of compiling and aggregating country wide input-output values to the specifications of this study, we adopted a more appropriate approach. First, representative values are used to reflect the coefficients for the two regions. For example, capital-output coefficients of India and the United States are used to represent the developing and developed countries respectively. To overcome the problem of aggregating large sector tables into a three sector table, we identified previous studies which have used approximately the same level of sectorial aggregation that we use in this study.

Once representative values were identified, we then used these values as reference points for the calibration process in our study. Tables 1 and 2 show the estimates which were used in the calibration process.

REGIONS	STUDY SOURCES		
	Kendrick and Taylor	Bagchi	Leontief
Agriculture	N.A	0.2	2.1
Industry	N.A	2.2	2.4
Services	N.A	1.47	1.8

Table 1 Capital-output coefficients for the developed region used by various sources

REGIONS	STUDY SOURCES		
	Kendrick and Taylor	Bagchi	Leontief
Agriculture	1.32	0.26	N.A
Industry	1.48	3.6	N.A
Services	1.35	1.58	N.A

Table 2 Capital-output coefficients for the developing region used by various sources

The Kendrick and Taylor (1969 pg 236) estimates are for South Korea in the 1960's and therefore reflect coefficients for the developing region. The coefficients from Leontief (Leontief 1986 pg 78) are indicative of the U.S. and are therefore taken to represent the developed region. The study by Bagchi (Bagchi 1984 pg 171) has estimates for both regions. The developing region coefficients are taken from the Indian economy while those of the developed region are those of the EEC countries[1].

The next step was then to solve the model with varying combinations of the capital-output coefficients while keeping the remaining parameters in the model constant. The important variables in this process were the GDP levels in the base year as well as the land-output coefficients. However, as mentioned in Chapter 4, the constant term in the land-output coefficients was computed without much difficulty. If the capital-output coefficients were not consistent with the remainder parameters of the variable, then no feasible solution could be found. The process was repeated until: (1) a feasible result was obtained; and (2) the solution was consistent with the desired paths. Tables 3 and 4 show the final capital-output coefficients which were adopted for this study.

Sectors	**Processes**		
	Pollution Intensive	Pollution Intermediate	Pollution Abatement
Agriculture	1.32	1.20	1.08
Industry	1.40	1.68	3.0
Service	1.25	1.5	2.5

Table 3. Capital-output coefficients for the developed region.

[1]These coefficient values are taken from the Bagchi study.

Sectors	Processes		
	Pollution Intensive	Pollution Intermediate	Pollution Abatement
Agriculture	1.46	1.33	1.20
Industry	2.11	2.74	4.0
Service	1.65	2.15	3.0

Table 4. Capital-output coefficients for the developing region.

The calibration process mentioned above was used to estimate the constant term in the capital-output coefficient. However, each process in the sectors has a constant term for its capital-output variable. We therefore make the assumption that the value which was calibrated above is indicative of an era which was not environmentally conscious and there reflect pollution intensive processes in the industry and service sectors. Once having identified the capital-output coefficient for the pollution intensive process in these two industries, we then projected the capital-output coefficients for the other two processes by assuming that the pollution intermediate and pollution abatement processes are more capital intensive by 50 and 100 percent, respectively (Ledbetter 1990).

The qualitative results of the experiments would not change if different values were used for the capital-output coefficients for the intermediate and abatement processes. Using lower values would have resulted in lower drops in consumption and GDP levels in the experiment but the same effect would have been observed in all the experiments. On the other hand, the coefficient values for the capital intensive process are sensitive to the experiments and the decision to swithch from a low to high CO_2 emitting process is determined to a large extent by these coefficient values. However, as these coefficients are derived from empirical observations, the confidence level of these values are significantly higher than the intermediate and abatement processes.

In the case on the agriculture sector, we assumed that the coefficient we calibrated is indicative of the pollution intensive process in the developed region but the pollution intermediate process in the developing region. This assumption was adopted on the basis of the degree of mechanization in the agriculture sectors in the two regions; the developed region's agriculture sector is highly mechanized and with high usage of fertilizers while the developing region has a lower level of mechanization and fertilizer use. This is supported by empirical data compiled by the FAO (FAO Yearbook 1985).

Land-Output Constant term

The land-output constant term was simpler to compute. We compiled the acreage under agriculture use and then divided the level of agriculture output by the acreage under tillage. The computation would have been more difficult if we had relaxed the assumption that land is only used by the agriculture sector. This will become an important factor in the future as the world's population increases and forest land is being cleared to make way for settlements and industries.

Tables 5 and 6 show the constant term values adopted for the land-output variable. As in the case of the capital-output coefficient, the computed values are used to represent a certain process within the sector. In the case of the developed region, the computed term is used to represent land use under the pollution intensive process. The computed term is used to represent land use under the pollution intermediate processes.

However, unlike the capital-output term in which the values of the other two processes were computed based on the base computation, we adopted a slightly different approach for the land-output term. We assumed that the pollution intensive process in the developing region can attain the value observed in the developed region and in the same way, the developed region's land usage under the pollution intermediate process will be similar to the developing region's. However, in the case of the pollution abatement process, we assume that the developed region finds it harder to adapt to a situation in which less mechanization is required and therefore has to use more land than that used in the developing region under the same conditions. The percentage increase in land usage under the pollution abatement process is assumed to be 40 percent greater than present use of pollution intermediate process. In the case of the developed region, it is assumed to require an increase of 250 percent over the present use of pollution intensive process.

Sectors	μ_{rsp}^{d} **for the Processes**		
	Pollution Intensive	Pollution Intermediate	Pollution Abatement
Agriculture	0.28	0.67	1.0
Industry	0.0	0.0	0.0
Service	0.0	0.0	0.0

Table 5 Land-output coefficients for the developed region.

Sectors	μ^d_{rsp} **for the Processes**		
	Pollution Intensive	Pollution Intermediate	Pollution Abatement
Agriculture	0.28	0.67	1.0
Industry	0.0	0.0	0.0
Service	0.0	0.0	0.0

Table 6 Land-output coefficients for the developing region.

At this point, we should like to point out that there are ample research possibilities in this area. However, as the focus of this study is on the modeling techniques and the advantages of using the holistic approach to economic-environmental issues, we believe that the data calibration and computation techniques used will satisfy these objectives. The difference in parameter estimates will in all probability change the quantitative results of the model but we believe that the qualitative results will hold.

BIBLIOGRAPHY

Abrahamson.D.E, ed. *Challenge of Global Warming.* Washington, D.C.: Island Press, 1989.

Arrhenius,S. On the Influence of Carbonic Acid in the Air upon the Temperature of the Ground, Phil Mag., 1896, 41, 237

Bagchi, Arunabha. *Stackleberg Differential Games in Economic Models.* Berlin: Springler-Verlag, 1984

Baumol, William J. and Wallace E. Oates. *The Theory of Environmental Policy* 2nd ed. 1975; rpt. Cambridge: Cambridge University Press, 1988

Bellman, R. Dynamic Programming. Princeton University Press, Princeton, New Jersey., 1957.

Biswas, A.K. Crop-Climate Models: A review of the state of the art, in Ausubel,J., and Biswas, A.K. (eds) Climatic Constraints and Human Activities, IIASA Proceeding Series, 1980, pp 75-92, Oxford, pergamon Press.

Bolin, Bert, et al. *The Greenhouse Effect Climatic Change and Ecosystems.* SCOPE 29. New York: John Wiley & Sons, 1986.

Bolin, B. (ed) Carbon Cycle Modeling, SCOPE 16, New York: John Wiley & Sons, 1981.

Brooke, Anthony; David Kendrick, and Alexander Meeraus. *GAMS:"A User's Guide".* Redwood City, California: The Scientific Press, 1988.

Brown,Lester, et al. *State of the World.* New York: W.W. Norton & Co. 1988.

Bryson, Arthur E. and Yu-Chi Ho. *Applied Optimal Control.* New-York: Hemisphere publishing Corp, 1975.

Central Intelligence Agency, *Relating Climate Change to its Effects*, GC78-10154, 1978.

Coase, R.H. "The Problem of Social Cost." Journal of Law & Economics, October 1960, pp 1-44.

Costa, A.M. *Development Planning: Techniques and Applications.* Amsterdam: Elsevier, North-Holland, 1981.

Daly, H.E. "Toward Some Operational Principles of Sustainable Development", Ecological Economics, Vol 2, No1, 1990.

Edmonds, J.A; et al, Future Atmospheric Carbon Dioxide Scenarios and Limitation Strategies, Noyes Publications, Park Ridge, New Jersey, 1986.

Edmonds,J., and Reilley,J. A long-term global energy-economic model of carbon dioxide release from fossil fuel use, Energy Econ., 5, 1983, pp 74-88.

Fujii, Y. CO2: A Balancing of Accounts. International Institute of Applied Systems Analysis (IIASA), Options, December 1990.

Fukui, H. Climatic variability and agriculture in tropical moist regions. In WMO, Proceedings of the World Climate Conference, WMO- No. 537, pp426-474, Geneva, 1979.

Gill, Philip; Walter Murray, and Margaret Wright. *Practical Optimization.* New York: Academic Press, 1981.

Green, Alex E.S., et al, *Greenhouse Mitagation*, The American Society of Mechanical Engineers, 1989.

Hafkamp, W.A., Economic-Environmental Modeling in a National-Regional System. Elsevier Science Publishers, Amsterdam, 1984.

Intriligator,M.D. *Mathematical Optimization and Economic Theory.* Englewood, N.j: Prentice-Hall, 1971.

Hammer, C.U., Clausen, H.B., and Dan sgaard, W. Greenland Icesheet Evidence of Post-Glacial Volcanism and its Climatic Impact, Nature, 1980, 288, 230-235.

Hansen, J. and Lebedeff, S. Journal of Geophysics, 1987, 92, 13345-13372.

Haun,J.R. *Mathematical Models in Agrometeorology,* CAgM Report No.14, Geneva, WMO, 1983.

Houghton, J.T; G.J. Jenkins, and J.J. Ephraums. *Climate Change: The IPCC Scientific Assessment.* Cambridge: Cambridge University Press, 1990.

Jaeger,J. "Impact Studies". *World Climate Programme*, WMO/TD -225, 1988.

Kendrick,David., and Lance Taylor. "numerical Solution of Non-Linear Planning Models". Econometricia, Vol 38. No3. May 1970, pg453-467.

Kendrick, David A. *Stochastic Control for Economic Models*, McGraw Hill Book Company, New-York, 1981.

Kendrick, David A. *Feedback: A New Framework for Macroeconomic Policy,* Kluwer Publishing Company, Dordrecht, Holland, 1988.

Kellog, W.W., and Robert Schware. *Climate Change and Society*, Boulder, Colorado: Wetview Press, 1981.

Kellog, W.W., Matthews, W.H, and C.D. Robinson. *Man's Impact on the Climate,* M.I.T Press, 1971.

Kerr, Richard A. "Global Warming Continues in 1989". Science, February 1990, pp521

Lashof, D.A., and Dennis A.Tirpak. *Policy Options for Stabilizing Global Climate.* New York: Hemisphere Publishing Corporation, 1990.

Legget, Jeremy, ed. *Global Warming The Greenpeace Report.* Oxford: Oxford University Press, 1990.

Leontief, W. *Input-Output Economics, Oxford:* Oxford University Press, 2nd Edition, 1986.

Leopold, Aldo. *A Sand County Almanac.* New York: Ballantine Books. 1970

MacCracken, M.C., and F.M. Luther. *Detecting the Climate Effects of Increasing CO2.* U.S. Department of Energy, DOE/ER-0235, December 1985.

Mackay, G.A., and T.Allsopp. *The role of Climate in affecting energy demand/supply.* Riedel, Netherlands: Dordrech, 1978.

Manabe, S., and R.T. Wetherald. "The effects of doubling the carbon dioxide concentration on the climate of a general circulation model". J. Atmos Sci. 32, 1975, pp.3-15.

Manabe. S., and R. Stouffer. "Study of climatic impact of carbon dioxide increase with a mathematical model of global climate". Nature. 282, 1979, pp491-493.

Manne,A., and R.G. Riechiels. "Global CO2 emission reductions--the impacts of rising energy costs", revised version of a paper presented to the International Association of Energy Economics, New Delhi; the INternational Institute for Applied Systems Analysis, Laxenburg; and Interaction Council, Amsterdam, (June), forthcoming in The Energy Journal (1990).

Manne, A. and R.G. Richels, "CO2 emission reductions: An Economic Cost Analysis for the U.S.A", The Energy Journal, Vol. 11. (1990)

Masters, G.M., *Introduction to Environmental Engineering and Science.* Prentice Hall, Englewood, New Jersey, 1991.

Nordhaus, W.D. "To Slow or not to Slow: The Economics of the Greenhouse Effect", revision of a paper presented to the 1989 meetings of the International Energy Workshop and the MIT Symposium on Environment and Energy, 1990.

Nordhaus,W.D., and Yohe,G. Future paths of energy and carbon dioxide emissions, Changing Climate, Washington, D.C., National Academy Press, 1983.

Nordhaus, W.D. The Efficient Use of Energy Resources, Cowles Foundation Monograph 26, Yale University Press, 1979.

Oeschger,H., Beer,J., Siegenthaler,U., and Stauffer, B. Late Glacial Climate History from Ice Cores. In Hansen, J.E., and Takahashi, T. (Eds). Climate Process and Climate Sensitivity (Maurice Ewing Series, No 5). American Geophysical Union, Washingtopn D.C. 1984, 299-306.

Parry,H.M. *Climate Change. Agriculture and Settlement.* Folkstone, England: Dawson, 1978.

Parry,M.L., and Carter,T.R. *The Effect of Climatic Variations on Agricultural Risk, Climate Change*, 7, 1985, pp 95-110.

Parry,M.L., and Carter,T. Assesing Impacts of Climatic Change in Marginal Areas: the Search for Appropriate Methodology, IIASA Working Paper WP-83-77, Laxenburg, Austria, International Institute for Applied Systems Analysis, 1984.

Parry,M.l., T.R. Carter, and N.T. Konijn. *The Impact of Climatic Variations on Agriculture, Volume 2: Assesments in Semi-Arid Regions,* Kluwer Academic Publishers, Dordrect, Holland, 1988.

Perry,H., and Hans H. Landsberg. Projected World Energy Consumption, Geophysics Study, 1987.

Ramanathan, V et. al "Climate and the Earth's Radiation Budget". Physics Today, May 1989, pp22-32.

Rasoul.S.I, ed. *Chemistry of the Lower Atmosphere.* New York: Plenun Press, 1973.

Rosenburg, N.J., and P.R. Carson. "A Methodology for Assessing Regional Economic Impacts of and Responses to Climate Change- The MINK Study". Resources fro the Future, Washington D.C. 1990.

Sellers, A.H., and McGuffie. K. *A Climate Modelling Primer,* John Wiley & Sons, 1987.

Siegenthaler,U., and Oeschger.H. "Predicting future atmospheric carbon dioxide levels". Science, 199, 1978, pp 388-395.

Thompson,L.M. "Weather Variability, Climate Change, and Grain Production". Science 189, 1978, pp535-541.

United Nations Environment Programme, The Greenhouse Gases, Nairobi, Kenya. UNEP Environmental Brief No ,1987.

United Nations Environment Programme, The Changing Atmosphere. Nairobi, Kenya. UNEP Environment Brief No1, 1988.

Varian, H.R. *Microeconomic Analysis*, 2nd ed. 1978; rpt. New York: W.W. Norton & Company, 1984.

Wigley, T.M.L. Impact of extreme events, Nature, 316, 1985, pp 106-107.

Wigley, T.M.L,. The Pre-Industrial Carbon Dioxide Level. Climatic Change, 1983, 5, 315-320.

Wigley, T.M.L., Ingram,M., and Farmer,G. (eds) *Climate and History: Studies in Past Climate and their Impact on Man*, Cambridge, Cambridge University press, 1981.

World Meteorological Organisation (WMO). *The Effect of Meteorological Factors on Crop Yields and Methods of Forecasting the Yield,* WMO No.566, Geneva 1982.

World Meteorological Organisation (WMO). The Reliability of Crop-climate Models for Assessing the Impacts of Climate Change and Variability. Report of the WMO/UNEP/ICSU-SCOPE expert meeting, Geneva, 1985.

Index

Advances in Computational Economics

1. A. Nagurney: *Network Economics.* A Variational Inequality Approach. 1993
ISBN 0-7923-9293-0
2. A. K. Duraiappah: *Global Warming and Economic Development.* A Holistic Approach to International Policy Co-operation and Co-ordination. 1993
ISBN 0-7923-2149-9

KLUWER ACADEMIC PUBLISHERS – DORDRECHT / BOSTON / LONDON

9 780792 321491

GPSR Compliance

The European Union's (EU) General Product Safety Regulation (GPSR) is a set of rules that requires consumer products to be safe and our obligations to ensure this.

If you have any concerns about our products, you can contact us on ProductSafety@springernature.com

In case Publisher is established outside the EU, the EU authorized representative is:

Springer Nature Customer Service Center GmbH
Europaplatz 3
69115 Heidelberg, Germany

Batch number: 10824484

Printed by Printforce, the Netherlands